Zoonotic Diseases and One Health

Zoonotic Diseases and One Health

Special Issue Editors

Marcello Otake Sato
Megumi Sato
Poom Adisakwattana
Ian Kendrich Fontanilla

MDPI • Basel • Beijing • Wuhan • Barcelona • Belgrade

Special Issue Editors

Marcello Otake Sato
Department of Tropical
Medicine and Parasitology,
Dokkyo Medical University
Japan

Megumi Sato
Graduate School of Health
Sciences, Niigata University
Japan

Poom Adisakwattana
Department of Helminthology,
Faculty of Tropical Medicine,
Mahidol University
Japan

Ian Kendrich Fontanilla
Institute of Biology,
College of Science,
University of the Philippines
Philippines

Editorial Office
MDPI
St. Alban-Anlage 66
4052 Basel, Switzerland

This is a reprint of articles from the Special Issue published online in the open access journal *Pathogens* (ISSN 2076-0817) in 2019 (available at: https://www.mdpi.com/journal/pathogens/special_issues/ Zoonotic_Diseases).

For citation purposes, cite each article independently as indicated on the article page online and as indicated below:

LastName, A.A.; LastName, B.B.; LastName, C.C. Article Title. *Journal Name* **Year**, *Article Number*, Page Range.

ISBN 978-3-03928-010-0 (Pbk)
ISBN 978-3-03928-011-7 (PDF)

Cover image courtesy of Marcello Otake Sato.

Contents

About the Special Issue Editors

Marcello Otake Sato, D.V.M. M.Sc. Ph.D. Doctor of Veterinary Medicine by the University of Sao Paulo, Brazil (Faculty of Veterinary Medicine) 1996. MSc in Biology (Biology of the Host–Pathogen Relationship), University of Sao Paulo, Brazil (Department of Parasitology, Institute of Biomedical Sciences) and Ph.D. in Medical Sciences (Zoonosis) at Asahikawa Medical University, Japan (Department of Parasitology).

Megumi Sato, Ph.D. B.Sc. in Health and Hygiene, University of the Air, Japan. M.Sc. in Tropical Medicine, Mahidol University (Department of Helminthology, Faculty of Tropical Medicine), Thailand. Ph.D. in Tropical Medicine, Mahidol University (Department of Helminthology, Faculty of Tropical Medicine), Thailand.

Poom Adisakwattana, Ph.D. B.Sc. (Medical Technology) Faculty of Allied Health Sciences, Thammasat University, Thailand. Ph.D. (Biomedical Sciences) Faculty of Allied Health Sciences, Thammasat University, Thailand.

Ian Kendrich Fontanilla, Ph.D. B.Sc. in Biology, University of the Philippines, Diliman. M.Sc. in Biology (Major in Genetics), University of the Philippines, Diliman. Ph.D. in Genetics, University of Nottingham, United Kingdom.

Preface to "Zoonotic Diseases and One Health"

One Health is a multidisciplinary and holistic approach with the perspective that the health of the environment—especially but not limited to animal health—is integral to public health. Zoonosis, or the spread of diseases between animals and humans, can be better understood and mitigated within the context of the shared environment. Understanding the mechanisms of transmission of zoonotic diseases within the different stages of their pathogens' lifecycles, the optimal environmental conditions for transmission, and even the effects of climate change on transmission can aid in the formulation of more appropriate policies and action plans towards a sustained One Health approach for public health.

This Special Issue of *Pathogens* highlights some recent works in selected countries that utilized the One Health approach, which recognizes the interconnections of the different components of the ecological communities and also includes the notion that humans are linked through interfaces with food, livestock, or exposure to the pathogens from the environment. Novel detection methods are likewise presented to better identify accurately unknown pathogens, their distribution, and the edaphic factors that contribute to their dispersal. Policies based on a multidisciplinary approach that address specific public health issues are also presented.

It is hoped that this Special Issue shall spur more studies towards greater understanding of the role of the environment in zoonotic transmission that will involve various stakeholders towards a truly One Health approach.

Marcello Otake Sato, Megumi Sato, Poom Adisakwattana, Ian Kendrich Fontanilla
Special Issue Editors

Review

Systematic Review of Important Bacterial Zoonoses in Africa in the Last Decade in Light of the 'One Health' Concept

Jonathan Asante [1], Ayman Noreddin [2] and Mohamed E. El Zowalaty [2,*] 

[1] Virology and Microbiology Research Group, School of Health Sciences, College of health Sciences, University of KwaZulu-Natal, Westville Campus, Durban 4000, South Africa; josante33@yahoo.com

[2] Infectious Diseases and Anti-Infective Therapy Research Group, Sharjah Medical Research Institute and College of Pharmacy, University of Sharjah, Sharjah 27272, United Arab Emirates; anoreddin@sharjah.ac.ae

* Correspondence: elzow001@gmail.com or elzow005@gmail.com; Tel.: +971-(56)-307-9774

Received: 23 March 2019; Accepted: 11 April 2019; Published: 16 April 2019

Abstract: Zoonoses present a major public health threat and are estimated to account for a substantial part of the infectious disease burden in low-income countries. The severity of zoonotic diseases is compounded by factors such as poverty, living in close contact with livestock and wildlife, immunosuppression as well as coinfection with other diseases. The interconnections between humans, animals and the environment are essential to understand the spread and subsequent containment of zoonoses. We searched three scientific databases for articles relevant to the epidemiology of bacterial zoonoses/zoonotic bacterial pathogens, including disease prevalence and control measures in humans and multiple animal species, in various African countries within the period from 2008 to 2018. The review identified 1966 articles, of which 58 studies in 29 countries met the quality criteria for data extraction. The prevalence of brucellosis, leptospirosis, Q fever ranged from 0–40%, 1.1–24% and 0.9–28.2%, respectively, depending on geographical location and even higher in suspected outbreak cases. Risk factors for human zoonotic infection included exposure to livestock and animal slaughters. Dietary factors linked with seropositivity were found to include consumption of raw milk and locally fermented milk products. It was found that zoonoses such as leptospirosis, brucellosis, Q fever and rickettsiosis among others are frequently under/misdiagnosed in febrile patients seeking treatment at healthcare centres, leading to overdiagnoses of more familiar febrile conditions such as malaria and typhoid fever. The interactions at the human–animal interface contribute substantially to zoonotic infections. Seroprevalence of the various zoonoses varies by geographic location and species. There is a need to build laboratory capacity and effective surveillance processes for timely and effective detection and control of zoonoses in Africa. A multifaceted 'One Health' approach to tackle zoonoses is critical in the fight against zoonotic diseases. The impacts of zoonoses include: (1) Humans are always in contact with animals including livestock and zoonoses are causing serious life-threatening infections in humans. Almost 75% of the recent major global disease outbreaks have a zoonotic origin. (2) Zoonoses are a global health challenge represented either by well-known or newly emerging zoonotic diseases. (3) Zoonoses are caused by all-known cellular (bacteria, fungi and parasites) and noncellular (viruses or prions) pathogens. (4) There are limited data on zoonotic diseases from Africa. The fact that human health and animal health are inextricably linked, global coordinated and well-established interdisciplinary research efforts are essential to successfully fight and reduce the health burden due to zoonoses. This critically requires integrated data from both humans and animals on zoonotic diseases.

Keywords: Zoonosis; livestock; bacteria; antimicrobial resistance; animals; Africa; antibiotics; One-health; epidemiology

---

1. Introduction

Zoonoses are infectious diseases caused by pathogens through the natural transmission between animals and man, directly (through agents such as saliva, blood, mucous and faeces) or indirectly (i.e., through environmental sources and vectors) [1]. Of all known human pathogens, including viruses, bacteria, fungi and parasites, an estimated 61% are regarded as zoonotic, with approximately 73% of emerging and re-emerging infections being considered as zoonoses [2]. Globally, it is estimated that 2.5 billion cases related to zoonotic infections are recorded yearly, resulting in 2.7 million deaths [3]. Zoonotic diseases account for 25% of the infectious disease burden in low-income countries, as poverty increases the risk for zoonotic diseases in communities where people are in close contact with livestock and wildlife [4,5]. The World Health Organization (WHO) estimated that, in 2010, there were 600 million cases of foodborne diseases, 350 million of which were caused by pathogenic bacteria [6]. A combined disease burden is imposed on people in poor areas such as tropical and subtropical Africa, where there is the likelihood of zoonotic diseases coinfection with other pathogenic or infectious diseases, such as malaria, tuberculosis and HIV. These associated factors may increase the severity of diseases and the susceptibility of individuals to infectious zoonotic agents, thus enhancing their spread at the community level [7]. Examples of bacterial zoonoses include anthrax, botulism, plague and tularemia, which are listed in category A warfare agents [8,9]. Bacterial zoonoses listed in category B agents include brucellosis, foodborne agents (*E. coli* O157:H7, salmonellosis and shigellosis), glanders, psittacosis, melioidosis, Q-fever, and typhus fever [9]. Zoonotic pathogens such as *Campylobacter*, *Salmonella*, *Listeria monocytogenes* and the Enterobacteriaceae family are frequently found in livestock (avian, bovine, caprine, equine, ovine and porcine) as well as in wild animals, pets and rodents, causing foodborne diseases. In immunocompromised populations, such as those with a high prevalence of HIV infection, the occurrence of zoonotic diseases is even higher. HIV infection, by depressing the immune systems leads to increased severity of symptoms of many zoonotic diseases and prolonged illness [1].

The absence of effective human monitoring and surveillance programs for zoonotic diseases coupled with limited laboratory capacities leads to a lack of clinical alertness, resulting in underdiagnoses and the subsequent mismanagement of these diseases. This further presents a challenge in detecting new and re-emerging pathogens early [10,11]. Zoonotic pathogens that tend to cause epidemics are usually given more attention regarding characterisation and policy-making than those that do not, despite the latter group having a major impact on rural communities [8].

The public health burden and socioeconomic effects of zoonotic diseases may vary according to geographical location, with a lack of data on disease burden in developing countries resulting in an underestimation of their impact [8].

Antimicrobial resistance has become a subject of global interest; especially as the use of antimicrobial agents continue to rise in both clinical and veterinary practices [12]. Microorganisms adapt to the effects of antimicrobial agents through numerous mechanisms, to enable them to survive in the presence of therapeutic concentrations of the antimicrobials. Thus, infections caused by pathogenic bacteria have become increasingly difficult to treat, due to the various antibiotic resistance mechanisms deployed by bacteria to evade the effects of antibiotics [12,13].

Humans, animals and the environment are interconnected in a complex and diversified manner. The interaction between humans, animals and the environment means that infections/resistance that originate in humans, animals, foods and farms will predictably lead to the spread of infection/resistant bacteria and/or resistance genes in the environment [13,14]. This dissemination of resistance may be facilitated by excreta coming into contact with soils as well as surface and ground water [14]. Thus, the 'One Health' approach seeks to amalgamate human and veterinary medicine, environmental sciences and public health to develop effective surveillance techniques, accompanied by appropriate diagnostic and therapeutic interventions. This holistic and coordinated approach will lead to the enactment of more thorough and effective policies [15].

This is the first timely, comprehensive, and updated systematic review about the significant bacterial zoonotic diseases in Africa over the past decade. The review summarises relevant publications reporting on occurrence, diagnosis and control of bacterial zoonoses in Africa within the last decade. The special focus of this study on Africa is explained by the limited data on disease burden of bacterial zoonoses within the continent, as well as the lack of effective monitoring and surveillance policies/techniques. The majority of African countries are classified as low- and middle-income nations; hence, the risk of disease transmission in communities in close contact with livestock is compounded by poverty. Furthermore, several countries in Africa specifically western and eastern Africa are at high risks of zoonotic diseases, where there are areas characterized by interplay of intense livestock animals, agricultural activities, and poor health services [16]. Furthermore, the risk of disease transmission in communities in close contact with livestock is compounded by poverty. Thus, the review provides important information to fill in the information gap.

2. Methods

2.1. Systematic Review Protocol

The systematic review followed the standard systematic review procedures established by the Preferred Reporting Items for Systematic Reviews and Meta-Analyses (PRISMA). The review used the following guidelines: (a) a database search to identify potentially relevant articles, (b) evaluating the relevance of articles, (c) quality assessment and (d) extraction of data, and are summarised in Figure 1.

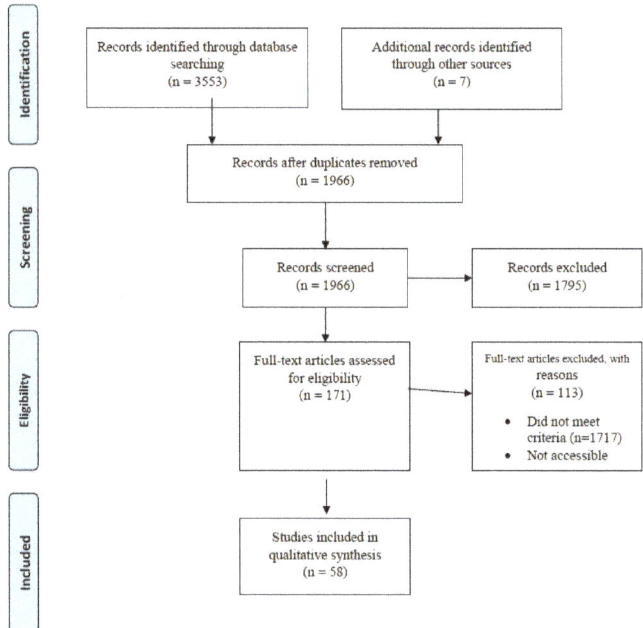

Figure 1. Preferred Reporting Items for Systematic Reviews and Meta-Analyses (PRISMA) flowchart showing search strategy and selection process for the research articles published between 2008 and 2018 used in the current study. Based on the search strategy, 3553 English articles were identified in total. Duplicates were removed.

2.2. Search Strategy and Data Collection/Extraction

In August 2018, we searched the English literature published between 2008 and 2018 on three scientific database search engines (PubMed, Web of science and Science Direct) for relevant articles using the search terms (Bacterial zoonoses OR zoonotic bacterial pathogens) AND (Africa) for articles published between January 2008 and August 2018. Other related articles that arose during the search, including bibliographies from selected papers were reviewed and added as additional information sources. Duplicate entries were identified and removed before the final selection of articles. Studies that did not meet the predetermined inclusion criteria were removed and included those outside the scope of Africa, nonbacterial zoonoses, conducted/published before 2008, non-English language, reviews, abstracts and conference proceedings. Citations were compiled and deduplicated using EndNote (Thomson Reuters, New York, NY, USA).

2.3. Data Screening

The full texts of retrieved articles were screened for inclusion. Studies were selected for evaluation if they met the following inclusion criteria.

- Any research article published between January 2008 and August 2018 that discusses bacterial zoonoses in Africa in both humans and animals.
- Any article that describes information relating to the occurrence (including outbreaks), diagnosis and control of bacterial zoonoses from any country, as defined by the United Nations (UN), within the stated period. Bacterial zoonoses/zoonotic bacterial pathogens were selected for inclusion based on the classification given by the individual studies.

Articles classified as eligible for inclusion were retrieved in full text format and were assessed using the case definitions specified by the respective studies (Table 1). Only accessible articles were screened. Studies were included if they reported on data from any country in Africa within the United Nations (UN) definition of Africa [17]. Only diseases/pathogens that routinely involve animal to human transmission were considered. Pathogens such as *Escherichia coli* and *Staphylococcus aureus*, which may or may not involve animal reservoirs, were excluded.

2.4. Data Analysis

The statistical analysis was carried out using SPSS version 25 [18] and R software version 3.5.2. [19].

Table 1. Case definitions in humans and animals.

Disease	Criteria			Reference
Brucellosis	**Confirmed**	**Probable**		
	Positive qPCR results or positive RBPT results confirmed by positive ELISA results			[20]
	Blood culture or a ≥4-fold increase in microagglutination test titre	Single reciprocal titre ≥160		[21]
	Presumptive acute brucellosis	**Probable prior brucellosis exposure**		
	Positive ELISA IgM antibodies result for *B. abortus*	Positive anti-*Brucella* IgG ELISA result		[22]
Q fever	**Acute Q fever**	**Chronic Q fever**	**Exposed**	
	Evidence criteria consistent with clinical evidence and supported by laboratory results indicated by elevated levels of ELISA IgG phase I and phase II antibodies and confirmed by IFA assay showing *C. burnetii* phase II antibodies titres of >1:128 or qPCR detection of *Coxiella* DNA	Cases with elevated ELISA IgG phase I antibodies and IFA assay phase I antibodies titres of ≥1:800.		[23]
	Clinical symptoms confirmed by qPCR targeting the IS1111 andIS30A spacers			[24]
	A ≥4-fold increase in immunoglobulin (Ig) G IFA titre to *Coxiella burnetii* phase II antigen		Titre ≥1000 to Phase I antigen or ≥64 to Phase II antigen on either sample defined Q fever exposure among those serum samples not meeting the case definition for acute Q fever	[25]
Spotted fever group rickettsiosis (SFGR) and typhus group rickettsiosis (TGR)	**Acute**		**Exposed**	
	A ≥4-fold increase in IgG IFA titre to *Rickettsia conorii* or *Rickettsia typhi* antigen		Titre to *R. conorii* or *R. typhi* ≥64 defined SFGR or TGR exposure, respectively, among samples that did not meet case definition for acute	[25]
Leptospirosis	**Acute**	**Presumptive acute leptospirosis**	**Probable prior leptospirosis exposure**	
	A MAT cut-off titre of ≥1:160	Positive IgM antibodies result *for Leptospira*	Positive anti-Leptospira IgG ELISA result	[22]
	Microagglutination test (MAT) > 400	IgM-positive/MAT < 400		[26]
	Confirmed *Leptospira* infection	**Probable leptospirosis**	**Exposure to pathogenic leptospires**	
	A ≥ four-fold increase in MAT titre	MAT titre ≥ 800	Titre ≥ 100	[27]
	Positive culture detection of Leptospira and/or positive PCR-specific assay for pathogenic *Leptospira* spp. Also, pathogenic serovar titre ≥ 200 considered positive by MAT			[28]
Plague	**Confirmed**	**Suspected**	**Probable**	
	clinically compatible acute illness with isolation of *Y. pestis* from a clinical specimen OR > 1 positive antibody titre against the F1 antigen of *Y. pestis*	Clinically compatible acute illness without laboratory confirmation	Suspected case linked epidemiologically to a confirmed case OR suspected case with further nonconfirmatory laboratory evidence of plague infection	[29]
Tularaemia	**Positive**	**Negative**		
	optical density > 0.25 (ELISA)	optical density <0.20 were considered negative		[30]

2.5. Quality Assessment and Data Extraction

Two independent researchers conducted full texts analysis of each publication using a data extraction form to extract predetermined qualitative and quantitative data; inconsistencies were decided by consensus. Data that consisted of sample size, infection prevalence, diagnosis/investigations, disease/pathogen, host/vector, country and year of study/publication were extracted from included eligible articles and compiled. The independent researchers examined eligibility of studies according the following criteria: appropriate description of study design which guaranteed the quality of the methodology, description of population and sample size for epidemiological studies and strength of association for studies reporting on risk for human infection. Articles were excluded if there was insufficient information in the methodology to decide if criteria were met. Studies that satisfied requirements for quality assessment were considered of enough quality to provide evidence of bacterial zoonoses in different host populations or probable predisposing risk factors.

2.6. Ethical Approval

This article does not contain any experimental studies involving human participants or animals performed by any of the authors. Parts of the manuscript involving data from ongoing research projects where ethical approvals were obtained from the Animal Research Ethics Committee of the University of KwaZulu-Natal (Reference: AREC 071/017 and AERC 014/018). The field sampling protocols, samples collected from animals and the research were conducted in full compliance with Section 20 of the Animal Diseases Act of 1984 (Act No 35 of 1984) and were approved by the South African Department of Agriculture, Forestry and Fisheries DAFF (Section 20 approval reference number 12/11/1/5 granted to Prof Dr. ME El Zowalaty).

3. Results

3.1. Data Acquisition

The preliminary database search yielded 3553 results. Manual search identified seven additional articles. Deduplication yielded 1966 unique articles. Reports were considered duplicated if they had the same information in the author, year of publication, name of the peer review, volume issue and page number fields. After removal of papers that did not meet the inclusion criteria, 58 papers were left for data extraction and qualitative analysis (Table 2). These included 15 articles reporting on *Brucella* spp. [20–22,31–42]; nine reporting on *Leptospira* spp. [22,26–28,43–47]; 13 reporting on *Coxiella burnetii* [23–25,39–41,48–54]; five on *Mycobacterium bovis* [42,55–58]; eight on *Rickettsia* spp. [25,53,54,59–63]; five reporting on *Anaplasma* spp. [53,63–66]; two each on *Bartonella* spp. [67,68] and *Borrelia* spp. [69,70]; one each reporting on *Yersinia pestis* [29], *Bacillus anthracis* [71], *Francisella tularensis* [30], *Ehrlichia canis* [53] and *Burkholderia pseudomallei* [40]; and six studies reporting on other zoonotic pathogens including *Salmonella* [72–75] and *Campylobacter* [76,77] (Table 2). Fourteen studies reported on human zoonoses, 33 were reports on animals, while 11 studies reported on both humans and animals (Table 2).

Table 2. Diagnoses, sources and study outcomes of bacterial zoonoses in Africa between 2008 and 2018.

Country	Period of Study	Year of Publication	Disease/Pathogen	Host/Vector/Source	Diagnostic Test/Investigations	Number of Animals/Humans/Samples Tested	Study Outcome/Disease Frequency/Seroprevalence	Reference
NORTHERN AFRICA								
Algeria	2011–2013	2016	Q fever (*Coxiella burnetii*)	Small ruminant flocks (aborted females)	Indirect ELISA, real time PCR (q-PCR)	494 samples (227 sera and 267 genital swabs)	*C. burnetii* seroprevalence was 14.1%. Bacterial excretion observed in 60% of flocks whiles 21.3% of females showed evidence of *C. burnetii* shedding.	[49]
Egypt	2008–2009	2014	Lyme borreliosis/*Borrelia burgdorferi*	Cattle, dogs, humans	Culture, PCR, enzyme-linked immunosorbent assay (ELISA)	92 samples (15 human blood samples, 25 cattle, 26 dog blood samples and 26 ticks)	24 out 77 non-human samples (51 blood and 26 tick) positive for the *OspA* gene. All human serum samples positive for IgM against *B. burgdorferi*	[69]
Egypt	2014	2014	*Brucella* spp.	Cattle, buffaloes	iELISA, qPCR	215 unpasteurised milk samples	34 (16%) samples were positive for anti-*Brucella* antibodies (iELISA) whiles qPCR detected *Brucella*-specific DNA from 17 (7.9%) milk samples.	[33]
Egypt	2015	2015	*Brucella abortus*	Cows, buffaloes, Egyptian Baladi goats and ewe	RBT, CFT, ELISA	25 serum samples from aborted animals	All 25 samples positive by PCR, but 10 positive by serology. *B. abortus* DNA was detected in all serum samples taken from buffaloes, goats, ewe and cows.	[35]
Egypt	2015	2015	Leptospirosis	270 rats, 168 dogs, 625 cows, 26 buffaloes, 99 sheep, 14 horses, 26 donkeys and 22 camels, humans and water sources	Culture, PCR and MAT.	Samples from 1250 animals, 175 human contacts and 45 water sources	Leptospira isolation rates were 6.9%, 11.3% and 1.1% for rats, dogs and cows, respectively. PCR detection rates were 24%, 11.3% and 1.1% for rats, dogs and cows, respectively.	[28]
Egypt	2016	2016	Bovine brucellosis (*Brucella abortus*)	cattle	Culture and biochemical tests, PCR, RBT, serum agglutination test (SAT), complement fixation test (CFT)	Samples selected from an outbreak in which 21 out of 197 pregnant, previously vaccinated cows aborted.	Two *B. abortus* biovar (bv.) 1 smooth and two *B. abortus* rough strains detected	[31]
Egypt	2015	2016	*Salmonella enterica* serovar Typhimurium	Chicken meat and humans	Culture, antimicrobial sensitivity testing, PCR.	700 samples (500 fresh chicken meat samples, 100 hand swab and stool samples each from workers)	Seventy-eight (11.1) of samples were *Salmonella* isolates, of which 18 were from humans and 60 from chicken samples). The virulence genes *stn*, *avr*A, mgtC, *inv*A and *bcf*C were detected in all screened isolates	[75]
Egypt	2017	2017	Q fever (*Coxiella burnetii*)	Small ruminants and humans	Serological assay	183 samples (109 sheep, 39 goats and 35 humans)	Seroprevalence of *C. burnetii* IgG antibodies were 25.71%, 28.20% and 25.68% in humans, goats and sheep, respectively	[50]
Egypt	2016	2017	Q fever (*Coxiella burnetii*)	27 sheep, 29 goats, 26 cattle, 26 buffaloes	Nested PCR, ELISA	108 aborted dairy animals, 56 human contacts	3.4% prevalence in goats, 0.9% overall prevalence, 19% prevalence in humans examined	[48]
Sudan	2007–2009	2013	Bovine tuberculosis	Cattle	Microscopy, culture, PCR	6680 bovine carcasses	Bovine TB infection rate was 0.18%.	[55]
Tunisia	2015	2017	*Anaplasma platys*-like infection	Goats, sheep and cattle	Restriction Enzyme Fragment Length Polymorphism (RFLP) assay, hemi-nested *groEL* PCR	963 domesticated ruminants	Prevalence rates were 22.8, 11 and 3.5% in goats, sheep, and cattle, respectively.	[65]

Table 2. *Cont.*

Country	Period of Study	Year of Publication	Disease/Pathogen	Host/Vector/Source	Diagnostic Test/Investigations	Number of Animals/Humans/Samples Tested	Study Outcome/Disease Frequency/Seroprevalence	Reference
WESTERN AFRICA								
Benin	2011	2016	Spotted fever group rickettsiae	*Amblyomma variegatum*	PCR.	910 ticks	Nearly one-third (29.4%) of samples (267/910) were positive for the SFG rickettsia-specific *ompA* gene, whereas 63.4% were positive by 16S rDNA gene amplification	[60]
Burkina Faso, Togo	2011–2012	2013	Brucellosis and Q Fever	Humans and livestock	RBT, ELISA, immunofluorescence assay (IFA)	683 people, 596 cattle, 465 sheep and 221 goats, 464 transhumant cattle from Burkina Faso	7 *Brucella* seropositive in humans, 9.2% seropositivity in village cattle, 7.3% in transhumant cattle and 0% in small ruminants	[41]
Côte d'Ivoire	2012-2014	2017	Brucellosis, Q Fever	Livestock and humans	Rose Bengal Test (RBT), indirect and competitive ELISAs for the respective pathogens	633 cattle, 622 small ruminants and 88 humans	Human seroprevalence for *Brucella* spp. was 5.3%., 4.6% seroprevalence in cattle adjusted for clustering. Q Fever seroprevalence was 13.9% in cattle, 9.4% in sheep and 12.4% in goats.	[39]
The Gambia	2014	2017	Q fever	Humans and small ruminants	ELISA, PCR	599 human serum and 615 small ruminant serum samples	24.9 seropositivity rate in small ruminants, and 3.8–9.7% in adults depending on ELISA test cut off	[51]
Ghana	2012	2012	*Bartonella* species	Bat flies	Culture and PCR analysis	137 adult flies	*Bartonella* DNA was found in 66.4% of specimen	[68]
Guinea	2011	2014	Brucellosis	Cattle	RBT, CFT	300 serum samples	29/300 RBT-positive, 26 of which were confirmed by CFT. Mean brucellosis prevalence for 2 communities was 8.67%.	[37]
Mali	2007–2011	2012	Tick-borne relapsing fever/ *Borrelia crocidurae*	*Ornithodoros sonrai* ticks, rodents and shrews.	Microscopy, serology (immunoblot)	663 rodents, 63 shrews and 278 ticks	Seroprevalence of *Borrelia* was 11.0% and 14.3% in rodents and shrews respectively	[70]
Niger	2009–2011	2015	Leptospirosis	*Arvicanthis niloticus, Cricetomys gambianus, Mastomys natalensis, Mus musculus* and *Rattus rattus*	qPCR, 16S-based metabarcoding, rrs gene sequencing, VNTR	578 samples	Leptospires not detected in *R. rattus* and *Mastomys natalensis*, but *Leptospira kirschneri* was detected in *Arvicanthis niloticus* and *Cricetomys gambianus*	[46]
Nigeria	2012	2014	Bovine tuberculosis	Cattle	Ziehl-Neelsen test, duplex PCR	168 lung samples	Prevalence of *Mycobacterium tuberculosis* was 21.4% (AFB test) and 16.7% (duplex PCR), 81.8% of lungs with lesions were positive whiles 6.7% of lungs without lesions were positive for AFB.	[58]
Nigeria	2012-2013	2014	*Bartonella* Species	Bats and Bat Flies	qPCR, DNA sequencing	148 bats and 34 bat flies samples	51.4% of bat blood samples and 41.7% of bat flies tested were positive for *Bartonella* spp. DNA. The prevalence by culture of *Bartonella* spp. among 5 bat species ranged from 0% to 45.5%.	[67]
Nigeria	2014	2015	Bovine tuberculosis (*Mycobacterium bovis*)	Cattle	PCR, Ziehl–Neelsen (ZN) staining	800 slaughtered cattle samples	120 samples classified as suspected bTB at postmortem, 29.2% and 8.3% of which were bTB-positive by ZN and PCR respectively	[56]
Nigeria	2007–2012	2016	Bovine tuberculosis	Cattle	N/A	52, 262 slaughtered cattle samples	11.2% showed signs of tuberculosis lesion at post mortem. Average yearly prevalence of bTB was 9.1%.	[57]
Nigeria	2011, 2015	2018	*Coxiella burnetii* and *Rickettsia conorii*	Rodents, fleas	PCR	194 peridomestic rodents, and 32 associated ectoparasites	2.1% of rodents carried *C. burnetii* DNA. All ectoparasites negative for *C. burnetii* by PCR, 6.3% of the pools of various ectoparasites were positive for *Rickettsia* spp. *gltA* PCR amplification	[54]
Senegal	2008–2009	2010	*Rickettsia felis*	Humans	qPCR	204 samples from 134 patients	Prevalence of spotted fever in all samples was 4.4% (9/204)	[61]

Table 2. *Cont.*

Country	Period of Study	Year of Publication	Disease/Pathogen	Host/Vector/Source	Diagnostic Test/Investigations	Number of Animals/Humans/Samples Tested	Study Outcome/Disease Frequency/Seroprevalence	Reference
EASTERN AFRICA								
Ethiopia	2007–2008	2011	Brucellosis	Cattle	RBT, CFT	1623 cattle sera	3.5% and 26.1% of animals and herds tested respectively had anti-*Brucella* antibodies.	[32]
Ethiopia	2011–2014	2015	Spotted fever group (SFG) rickettsiae	Ixodid ticks collected from domestic animals	Quantitative PCR (qPCR) system targeting the *glt*A gene	767 ixodid ticks	*Rickettsia africae* DNA was detected in 30.2% of Amblyommma variegatum, 28.6% *Am. gemma*, 0.8% *Am. cohaerens*	[59]
Ethiopia	2013	2016	Salmonellosis/*Salmonella* spp.	Dairy cattle	Culture, biochemical tests, PCR, antimicrobial susceptibility testing, serotyping and phage typing	1203 faecal samples	30 samples positive for *Salmonella*. Standard serological agglutination tests identify 9 different serotypes, with *Salmonella typhimurium* (23.3 %) being the most dominant	[73]
Ethiopia	2015	2017	Salmonellosis/*Salmonella* spp.	Dogs	Culture, antimicrobial susceptibility testing, serotyping and phage typing	360 dogs	42 (11.7%) *Salmonella*-positive. 14 serotypes detected	[74]
Kenya	2009	2010	Salmonellosis/*Salmonella* spp.	Pigs	Biochemical tests, serotyping, phage typing and PCR	116 samples	13.8% positive for *Salmonella*, 35.7% of isolates displayed antimicrobial resistance, 7.1% displayed multidrug resistance	[72]
Kenya	2012–2013	2015	Brucellosis	Humans and animals (cattle, sheep, camels, and goats)	ELISA	1088 households surveyed. 11,028 livestock (37% goats, 28% sheep, 27% cattle, and 8% camels) were sampled	Individual human and animal seroprevalence were 16 and 8% respectively. Household and herd prevalence ranged from 5–73%, and 6–68%, respectively	[38]
Kenya	2014–2015	2016	Brucellosis	Humans	Modified Rose Bengal Plate Test (RBPT), ELISA, PCR.	1067 patients	146/1067 (13.7%) tested positive for brucellosis. *B. abortus* the only *Brucella* species found using species-specific qPCR	[20]
Kenya	2014–2015	2016	Q fever	Humans	ELISA, IFA, qPCR	1067 patients	19.1% of sera were seropositive by qPCR. 16.2% of patients had acute Q fever.	[23]
Kenya	2016	2016	Q fever	Humans and cattle	ELISA	2049 human serum and 955 cattle serum samples	Overall seroprevalence of *Coxiella burnetii* was 10.5% in cattle and 2.5% in humans	[52]
Kenya	2013–2014	2017	Novel Rickettsia	Adult ticks, nymphs and larvae	PCR	4297 questing ticks	*Anaplasma phagocytophilum* detected in *Rh. maculatus* ticks and a first-time detection of *Ehrlichia chaffeensis*, *Coxiella* sp., *Rickettsia africae* and *Theileria velifera* in *Am. eburneum* ticks	[62]
Kenya	2014–2015	2017	Tularaemia (*Francisella tularensis*)	Humans	ELISA and Western blot	730 patients	71 (9.7%) were seropositive for *F. tularensis* by ELISA but 27 (3.7%) were confirmed by Western blotting	[30]
Madagascar	2010–2012	2014	*Leptospira*	Small mammals	PCR	344 samples	44 samples (12.8%) positive for *Leptospira* spp.	[44]
Madagascar	2011–2013,	2017	Brucellosis (*Brucella* spp.), Q fever (*Coxiella burnetii*) and melioidosis (*Burkholderia pseudomallei*)	Human, cattle and ticks	Specific quantitative real-time PCR assays (qPCRs)	1020 blood samples from febrile patients, 201 Zebu cattle serum samples and 330 zebu cattle-associated ticks	15 (1.5%) of samples were *Brucella*-positive, and 0% for *C. burnetii* and *Bu. Pseudomallei*. Anti-*C. burnetii* antibodies detected in 4 zebu serum samples, but no anti-*Brucella* antibodies were detected, 1% of ticks analysed tested positive for *C. burnetii* DNA.	[40]
Madagascar, Union of the Comoros	2012	2012	*Leptospira* spp.	Bats	qPCR	129 bats (52 from Madagascar and 77 from Union of the Comoros)	25 samples were positive by probe-specific qPCR. There were 34.6% and 11.7% infection rates in bats from Madagascar and Comoros, respectively.	[43]

Table 2. *Cont.*

Country	Period of Study	Year of Publication	Disease/Pathogen	Host/Vector/Source	Diagnostic Test/Investigations	Number of Animals/Humans/Samples Tested	Study Outcome/Disease Frequency/Seroprevalence	Reference
Malawi	2011	2014	Brucellosis and bovine tuberculosis (bTB)	Cattle	Competitive ELISA, tuberculin skin test	156 and 95 cattle respectively tested for brucellosis and bTB	7.7% and 1.1% of the 156 and 95 cattle respectively tested positive for brucellosis and bTB	[42]
Mozambique	2012–2015	2017	Leptospirosis	Humans	ELISA, microagglutination test (MAT)	373 paired serum samples from febrile patients	1.3% had acute leptospirosis (MAT > 400), 10.2% had a presumptive infection (IgM-positive/MAT <400).	[26]
Tanzania	2007–2008	2011	Leptospirosis	Humans	MAT, blood culture	870 patients	8.8% of 453 paired (acute and convalescent) sera samples were confirmed leptospirosis, 3.6% of 832 patients (with ≥ 1 serum sample available) classified as having probable leptospirosis.	[27]
Tanzania	2007–2008	2011	Q Fever, Rickettsioses (Spotted Fever Group, SFGR and Typhus Group, TGR)	Humans	ELISA, culture	870 patients, 483 tested for acute Q fever, 450 tested for acuteSFGR and TGR	Infection rates of acute Q fever, SFGR and TGR were 5.0%, 8.0% and 0.5% respectively.	[25]
Tanzania	2007–2008	2012	Brucellosis	Humans	Blood culture, MAT	870 patients	455 (52.3%) had paired sera available. 16/455 (3.5%) were confirmed brucellosis, 830 people had ≥ 1 serum sample of which 0.5% had probable brucellosis	[21]
Tanzania	2013	2015	Leptospirosis, brucellosis	Humans	MAT, IgM and IgG ELISA	370 patients	11.6% had presumptive acute leptospirosis, whiles 7.0% and 15.4% showed presumptive acute brucellosis due to *B. abortus* and *B. melitensis*, respectively.	[22]
Tanzania	2011–2012	2015	*Campylobacter*	Humans	Culture, matrix-assisted laser desorption/ionisation–time-of-flight (MALDI-TOF) mass spectrometry and PCR	1195 persons	11.4% *Campylobacter*-positive. *C. jejuni* (84.6%) was most abundant *Campylobacter* species, with *C. coli* being 15.4%.	[76]
Tanzania	2012–2014	2018	Leptospirosis	Rodents, cattle, goats, sheep	qPCR, culture, phylogenetic analysis	452 cattle, 167 goats, 89 sheep	7.08% of cattle, 1.20% of goats and 1.12% of sheep carried pathogenic Leptospira infection. No pathogenic Leptospira infection was found in rodent species sampled	[47]
Uganda	2014	2014	Brucellosis	Humans	Rapid Plate Agglutination Test, Standard Tube Agglutination Test (STAT), cELISA	329 individuals (161 exposed cattle keepers and 168 individuals attending HIV testing).	Brucellosis seroprevalence in exposed cattle keepers and consumers of raw milk were 5.8% and 9%, respectively.	[36]
Uganda	2012–2013	2016	Brucellosis	Pigs	ELISA, CFT	1665 serum samples	3 samples *Brucella*-positive by ELISA, which were in turn *Brucella*-negative by CFT. SAT detected anti-*Yersinia* enterocolitica antibodies in 2 samples	[34]
Uganda	2008–2016	2017	Plague (*Yersinia pestis*)	Humans	Culture, bacteriophage lysis	255 suspected cases	78 (31%) as confirmed per specified criteria	[29]
Zambia	2011	2012	Anthrax	Humans, hippopotamuses	Culture, PCR	56 samples from human patients, hippopotamuses and soil.	30.4% of samples were culture-positive. All isolates tested were resistant to vancomycin while isolates showed 100% susceptibility to mostly the penicillins	[71]

Table 2. *Cont.*

Country	Period of Study	Year of Publication	Disease/Pathogen	Host/Vector/Source	Diagnostic Test/Investigations	Number of Animals/Humans/Samples Tested	Study Outcome/Disease Frequency/Seroprevalence	Reference
Zambia	2008	2014	*Anaplasma phagocytophilum*, *Rickettsia* spp.	Yellow baboons (*Papio cynocephalus*) and vervet monkeys (*Chlorocebus Pygerythrus*)	PCR	88 spleen DNA samples	*Anaplasma phagocytophilum* and *Rickettsia* spp. were detected in 12 (13.6%) and 35 samples (39.8%) respectively.	[63]
Zambia	2016	2018	Anaplasmosis (*Anaplasma platys*)	Dogs	PCR	301 blood samples	9% prevalence of *Anaplasma* species	[64]
Zimbabwe	2014	2014	*Anaplasma phagocytophilum*.	Lions (*Panthera leo*), Southern African wildcats, cheetahs (*Acinonyx Jubatus*) and servals	PCR	98 whole blood samples from 86 lions, 6 Southern African wildcats, 4 cheetahs and 2 servals.	Mixed infection of *A. phagocytophilum* with other parasitic pathogens observed in 1 serval and 1 Southern African wildcat.	[66]
SOUTHERN AFRICA								
Botswana	2009–2012	2014	Leptospirosis (*Leptospira interrogans*)	Banded mongoose (*Mungos mungo*), Selous' mongoose (*Paracynictis selousi*)	PCR	42 samples (41 banded mongooses and 1 Selous' mongoose	41.5% prevalence among banded mongoose, the one Selous' mongoose sample was Leptospira-positive	[45]
Botswana	2017	2018	*Campylobacter* spp.	Humans, chickens	Culture, whole genome sequencing	20 human samples, 70 chicken samples	Phylogenetic analysis showed a high a level of relatedness between *Campylobacter* isolated from human and various poultry sources. Resistance determinants found include *tetO* (52%), *gyrA*-T86I (47%) and *bla*$_{OXA-61}$ (72%)	[77]
South Africa	Not stated	2017	*Coxiella burnetii*, *Ehrlichia canis*, *Rickettsia* species and *Anaplasma phagocytophilum*-like bacterium	*Rhipicephalus sanguineus*, *Haemaphysalis elliptica* and *Amblyomma hebraeum*	PCR	318 ticks from dogs and cats. 147 pooled DNA samples	Prevalence were 37% (*Rickettsia* spp.), 41% (*Coxiella burnetii*), 18% (*Ehrlichia* or *Anaplasma*), 18% (*Anaplasma phagocytophilum*-like bacterium) from pooled DNA samples	[53]
STUDIES IN DIFFERENT GEOGRAPHIC LOCATIONS								
Senegal, Mali, Tunisia, Algeria, Gabon, and Morocco	2008, 2010–2012	2014	Q fever (*Coxiella burnetii*)	Humans	qPCR to amplify the *IS1111* and *IS30A* spacers	1888 febrile patients (1238 from Senegal, 100 from Mali, 50 from Gabon, 184 from Tunisia, 268 from Algeria and 48 from Morocco), 500 nonfebrile samples	0.3% *C. burnetii* infection rate in Algeria, 0.5% in Senegal. No infection detected in Mali, Morocco, Gabon, and Tunisia. All nonfebrile samples were negative.	[24]

These articles reported on the occurrence, diagnostic methods of zoonoses in humans, livestock, companion animals and vectors. The studies varied in terms of methodological designs, sampling methods, sample size and diagnostic criteria. Most of the studies were case reports, while three were outbreak reports [29,31,71]. The risk of sampling bias in retrospective seroepidemiological studies may be significant considering that these studies utilised samples collected or submitted to research laboratories and thus did not provide evidence of random sampling.

The prevalence of different bacterial zoonotic diseases in the four geographic regions in Africa is shown in Figure 2. Bartonellosis was the highest prevalent disease (57.73%) in western Africa and leptospirosis was the highest prevalent (31.17) disease in northern Africa, plague was the highest prevalent (30.59%) in eastern region, while rickettsiosis was the highest prevalent (37%) in southern Africa.

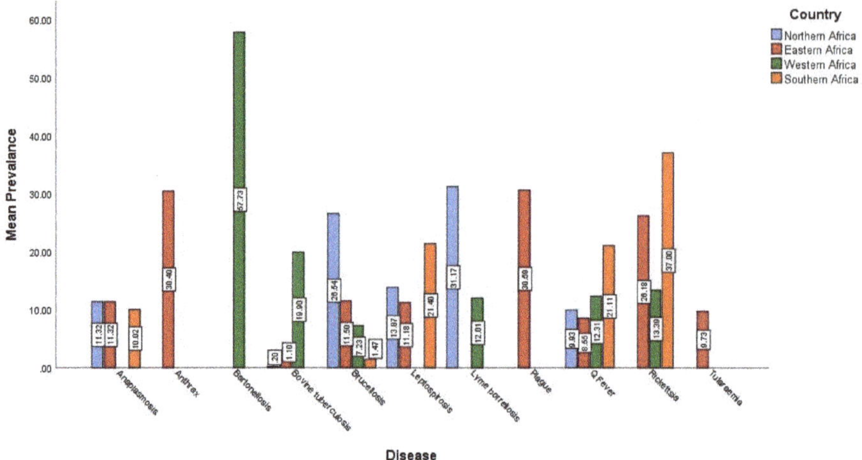

Figure 2. The prevalence of important bacterial zoonotic diseases in different geographic regions in Africa.

As shown in Figure 3, a map of Africa showed the location of the different studies by pathogen in different countries. There was no study that met the inclusion criteria reporting bacterial zoonotic diseases that in central Africa at the time of this review.

Figure 3. Geographic distribution of important bacterial zoonotic diseases between 2008 to 2018 in Africa. Map of Africa showing locations indicating countries with reported zoonotic diseases and circulation. (Map was reproduced from Nations Online Project.)

3.2. Brucellosis

Egypt was the most frequently represented country followed by Kenya and Uganda. The Rose Bengal Test (RBT), complement fixation test (CFT) and enzyme-linked immunosorbent assay (ELISA) were the main diagnostic tests used. Others included culture and biochemical tests, real time PCR (qPCR) and standard microagglutination test (MAT). The prevalence of brucellosis in humans was investigated by four studies including two hospital-based studies [20,21] and two in high risk occupational/population groups [36,39]. Njeru et al. (2016) sought to determine the prevalence of brucellosis in patients in two hospitals in Kenya and to define their clinical characteristics to help clinicians identify cases of brucellosis in regions with limited laboratory capacities. It was reported that 13.7% of samples tested were positive for brucellosis (defined as positive qPCR results or positive RBPT results confirmed by positive ELISA results) [20]. Bouley et al. (2012) also found evidence of brucellosis in 3.5 % of participants screened. There was no diagnosis of brucellosis by the hospital clinical team even though study participants with brucellosis were given antibiotics or antimalarials in the hospital [21]. Using blood samples, Boone et al. (2017), investigated the causes of febrile illness in Madagascar, and found a 1.5% detection rate for *Brucella* [40]. It is the first report of brucellosis in febrile patients reported in Madagascar [40]. Chipwaza et al. (2015) investigated the prevalence of bacterial febrile illnesses in Tanzania, and found that 7.0% and 15.4% showed presumptive acute brucellosis due to *B. abortus* and *B. melitensis*, respectively [22].

In the Uganda study of cattle keepers and consumers of unpasteurised milk, consumption of unpasteurised milk was significantly linked ($p = 0.004$) to seropositivity in one of the districts of the study (Mbarara District). Brucellosis seroprevalence in exposed cattle keepers and consumers of raw milk were 5.8% and 9%, respectively, in this study [36] (Table 2).

Six articles investigated brucellosis in animals including livestock [31–35,37]. In an outbreak investigation in Egypt, one study investigated the molecular profile of *Brucella* isolates and found two different profiles of the *B. abortus* biovar (bv.): one smooth and one rough *B. abortus* strain, with low genetic diversity identified by the molecular typing method and multiple locus of variable number tandem repeats analysis (MLVA-16) [31]. As risk factors for *Brucella* infection, Megersa et al. (2011) found that herd size and age of cattle were found to have played roles in a study investigating the prevalence of cattle brucellosis in traditional animal husbandry practice [32].

Large (odd ratio (OR) = 8.0, 95% CI = 1.9, 33.6) and medium herds (OR = 8.1, 95% CI = 1.9, 34.2) were found to present a higher risk of infection than small herds. One article investigated the prevalence and risk factors for brucellosis in humans and livestock, and found their individual seroprevalence to be 16% and 8%, respectively [38]. Risk factors found to affect the odds for human seropositivity in this study included exposure to goats (adjusted odds ratio (OR) = 3.1, 95% CI = 2.5–3.8), frequent consumption of raw milk (OR = 3.5, 95% CI = 2.8–4.4) and handling of animal hide (OR = 1.8, 95% CI = 1.5–2.2). Again, there was an association between seropositivity in humans and animals, with a six-fold increase observed for humans in households with seropositive animals compared to those without [38].

3.3. Q Fever

Three papers investigated the presence of Q fever in human febrile patients [23–25]. The study by Angelakis and colleagues (2014) was conducted in five countries—Senegal, Mali, Tunisia, Algeria, Gabon and Morocco—and recorded infection rates of 0.3% and 0.5% in Algeria and Senegal, respectively. For the first time in humans, *Coxiella burnetii* (causative agent of Q fever) genotype 35 was found in a patient in Senegal [24]. In the other study in febrile patients, 16.2% of patients screened had acute Q fever [23] (Table 1). Risk factors for human infection included exposure to goats (OR: 3.74, 95 % CI: 2.52–9.40), cattle (OR: 2.09, 95% CI: 1.73–5.98) and animal slaughters (OR: 1.78, 95% CI: 1.09–2.91). Dietary factors linked with seropositivity were found to include consumption of raw milk (OR: 2.49, 95% CI: 1.48–4.21) and locally fermented milk products (OR: 1.66, 95% CI: 1.19–4.37). Univariate analyses showed no significant association between county of residence, gender, occupations (except herders) and seropositivity. Using ELISA and culture assays, Prabhu et al. (2011) investigated the occurrence of Q Fever in hospitalised febrile patients in northern Tanzania, and found the infection rate to be 5.0% [25].

Five articles probed the presence of Q fever in human and animal hosts [41,48,50–52]. Abdel-Moein and Hamza examined vaginal discharges and placental cotyledons from animals that had aborted and found an overall prevalence of 0.9%, with the highest prevalence of Q fever being found in goats (3.4%). A seroprevalence of 19% was detected in the human contacts screened, with a higher prevalence being detected in farmers (30.6%) than veterinarians and veterinary assistants (9.4%) [48]. A higher seroprevalence of 25.71% was found in human contacts in Egypt [50].

In a Gambia study, a 24.9% seropositivity rate in small ruminants and 3.8–9.7% in adults, depending on the ELISA test cut off, was reported [51]. Having at least one seropositive animal in a compound was determined to be a risk factor for human seropositivity (OR: 3.35, 95% CI: 1.09–14.44) [51]. Wardrop et al. (2016) found overall *C. burnetii* seroprevalence in cattle and humans to be 10.5% and 2.5%, respectively [52]. There was no correlation between cattle and human seroprevalence. An article investigated the prevalence of Q fever infection in small ruminants after abortion or the lambing period and found a 14.1% prevalence at individual level and 58.6% at flock level in Algeria [49]. Excretion of bacteria was found in 60% of flocks, with 21.3% of females showing evidence of *C. burnetii* shedding. Dean and colleagues investigated the seroprevalence of Q fever in humans and livestock in Togo, and found that there was a significantly higher *C. burnetii* seroprevalence among the Fulani people, who also had greater livestock contact (45.5%, 95% CI: 37.7–53.6%) [41].

Real-time PCR (qPCR) and ELISA were the most commonly used diagnostic tests. Another test included indirect immunofluorescence assay (IFA) (Table 2).

3.4. Leptospirosis

PCR was the most widely used diagnostic method, being used in six out of the eight studies. Other techniques such as culture isolation, MAT and ELISA were also used. Three articles [22,26,27] studied the seroepidemiology of leptospirosis among febrile patients. In a Morocco study, Ribeiro et al. (2017) observed that 1.3% of samples had acute leptospirosis defined therein as a microagglutination test (MAT) > 400, while 10.2% had a presumptive infection, therein defined as IgM-positive/MAT <400. Patients with acute infection had a significantly higher contact with rodents (100%, 5/5) than those with presumptive (39.5%, 15/38) or no infection (41.8%, 138/330) ($p = 0.031$). Although the malaria tests proved negative, 80% of patients with acute leptospirosis were given antimalarial drugs. In addition, 20.9% of the confirmed/presumptive cases of leptospirosis occurred in sub-urban populations. Similarly, Biggs et al. (2011), in their study of leptospirosis in febrile patients in northern Tanzania, observed that 8.8% of paired (acute and convalescent) sera samples were confirmed leptospirosis (defined therein as ≥ four-fold increase in MAT titre) and 3.6% (with ≥1 serum sample available) were classified as having probable leptospirosis (defined therein as MAT titre ≥ 800). The most predominant serotypes were Mini and Australis. There was an association found between Leptospira infection and rural dwelling (OR 3.4, $p < 0.001$) [27]. Chipwaza et al. (2015) found 11.6% seroprevalence of presumptive acute leptospirosis among people presenting with febrile illnesses [22].

In a study in Egypt, Leptospira isolation rates were 1.1%, 6.9% and 11.3% for cows, rats and dogs, respectively, whereas PCR detection rates were 1.1%, 24% and 11.3%, respectively [28]. The human contacts who were tested proved negative by culture isolation and PCR. However, using MAT, the seroprevalence of the human samples was determined to be 49.7%. In that study, six Leptospira serovars (Grippotyphosa Pyrogenes, Icterohaemorrhagiae, Canicola, Celledoni and Pomona) were isolated from cows, rats and dogs. These three species of animals were found in this study to be the most important carriers of leptospirosis in Egypt. Of note is the recovery of some isolates from rats caught from dairy farms and water sources supplying the farms [28]. In a survey of an area with a high reported incidence of human leptospirosis in northern Tanzania, Allan et al. (2018) found no proof of Leptospira in rodents sampled randomly in and around households in the area. However, 7.08% of cattle, 1.20% of goats and 1.12% of sheep from local slaughterhouses carried pathogenic Leptospira infection [47]. Similarly, although *Rattus rattus* and *Mastomys natalensis* are usual rodent reservoirs for Leptospira, Leptopires was not detected in them, although *Leptospira kirschneri* was detected in two rodent species, namely, *Arvicanthis niloticus* and *Cricetomys gambianus*, which are confined to irrigated cultures in the city [46]. The variable number of tandem repeat (VNTR) profiles showed that the leptospires found did not belong to any previously described serovars. The first published report of *L. interrogans* in the Banded mongoose (*Mungos mungo*) and Selous' mongoose (*Paracynictis selousi*), and the only published report of the pathogen in wildlife in Botswana was reported by Jobbins et al. (2014) [45]. In some cases, the prevalence of Leptospira in animals including bats and other small mammals ranged from 11.7% to 34.6% [43,44].

3.5. Bovine Tuberculosis

A bovine tuberculosis infection rate of 0.18% was detected in a Sudan study, with prevalence of 4.5% in slaughtered cattle with caseous lesions [55]. Sa'idu et al. (2014) conducted a study to establish the prevalence of bovine tuberculosis in slaughtered cattle using PCR and Ziehl-Neelsen (ZN) staining and found an overall prevalence rate of 8.3% [56]. In study of bovine tuberculosis in slaughtered cattle in Nigeria, the prevalence of mycobacterium TB was 21.4% (acid-fast bacilli test) and 16.7% (duplex PCR) [58]. The presence of lesions in lungs was highly associated (OR = 52.3; 95% CI: 16.4–191.8) with positive results for acid-fast bacilli (AFB) test compared to those without lesions. A retrospective study at a Nigerian abattoir was conducted with an average yearly bovine tuberculosis prevalence rate of 9.1% detected [57].

3.6. Rickettsiosis

Four articles investigated *Rickettsia* spp. in ticks [53,59,60,62] and two in humans [25,61]. Prabhu et al. (2011) investigated the occurrence of spotted fever group (SFGR) and typhus group rickettsioses (TGR) in hospitalised febrile patients in northern Tanzania, and found infection rates to be 5.0%, 8.0% and 0.5%, respectively [25]. Kumsa et al. (2015) investigated the transmission of spotted fever group rickettsiae through ixodid ticks and found an overall prevalence to be 6%. Being the first study to investigate SFG rickettsiae in Benin, Moumouni et al. (2016) found that 29.4% of samples were positive for the SFG rickettsia-specific *ompA* gene, whereas 63.4% were positive by 16S rDNA gene amplification [60]. In Senegal, a study sought to investigate the cause of reported febrile conditions that had tested negative for malaria [61]. The prevalence of spotted fever in all samples was 4.4%, with was no positive sample recorded for typhus group rickettsiae. By sequencing theamplicons, one sample was found to be *R. conorii* [61].

3.7. Anaplasmosis

Vlahakis et al. (2018) conducted a study to identify and characterize *Anaplasma* species from dogs in Zambia and found a 9% prevalence of *Anaplasma* spp. as detected by PCR. It is the first study to highlight the prevalence of *Anaplasma* spp. in dogs in Zambia and the first report of *Anaplasma platys* in Zambia [64]. Said et al. (2017) used a restriction enzyme fragment length polymorphism (RFLP) together with a hemi-nested *groEL* PCR method to distinguish between *A. platys* and genetically related strains. Analysis of the sequence variants pointed to infection with an unclassified *Anaplasma platys*-like strains that were genetically related to *A. platys*, with prevalence rates ranging from 3.5% to 22.8% in sheep, goats and cattle [65]. Mtshali and colleagues identified an *Anaplasma phagocytophilum*-like bacterium in 18% of pooled DNA samples [53].

3.8. Lyme Borreliosis

Elhelw et al. (2014) investigated the occurrence of borreliosis as an emerging zoonotic disease and its zoonotic potential in Egypt [69] and found *Borrelia burgdorferi* in the animals screened. In addition, the *OspA* gene (outer surface protein A gene) and anti-*B. burgdorferi* IgM were detected by PCR and ELISA respectively in human contacts. The use of culture techniques to isolate *B. burgdorferi* showed low sensitivity as shown by the recovery of only one isolate out of seven samples cultured, while 26.6% of febrile human blood samples tested were positive by PCR, and 15 out of 15 serum samples tested positive for IgM ELISA. The human contacts had been exposed to tick bites, which suggests a possible zoonotic transfer. In Mali, Borrelia seroprevalence of 11.0% and 14.3% in rodents and shrews, respectively, was observed, with 2.2% of animals displaying active spirochete infections at the time of capture [70].

3.9. Bartonellosis

In a first report on the occurrence of *Bartonella* spp. in bats and bat flies from Nigeria, 51.4% of bat blood samples and 41.7% of bat flies tested were positive for *Bartonella* spp. DNA [67]. The prevalence by culture of *Bartonella* spp. among five bat species ranged from 0% to 45.5% [67]. Of 137 adult bat flies studied in Ghana, 66.4% were positive for Bartonella DNA [68].

3.10. Plague

In a suspected plague outbreak in Uganda, 31% (78 out of 255 suspected cases) of cases were confirmed as plague [29]. The study found a correlation between reports of human plague and a large number of dead rats in a village. Close contacts with rodents, lack of appropriate antibiotics and a delay in seeking medical help contributed to the menace of human plague in the area where the study was conducted [29].

3.11. Tularaemia

Among febrile patients seeking treatment at remote hospitals in northeastern Kenya, 9.7% were seropositive for *Francisella tularensis* by ELISA, while 3.7% were confirmed by Western blotting [30]. Most of the febrile cases that tested positive to tularaemia were not recognised by clinicians and the appropriate treatment protocol was not therefore followed. Indeed, most cases were treated with antimalarial agents and/or beta-lactam antibiotics.

3.12. Anthrax

In the light of a suspected outbreak of anthrax in Zambia in 2011, a study to investigate the cause was initiated [71]. Human, hippopotamus and soil samples were screened by culture and PCR methods. It was found that 30.4% of samples were culture-positive. All isolates tested were resistant to vancomycin, but showed 100% susceptibility to the penicillins [71].

3.13. Others

3.13.1. Salmonella

In a study probing antimicrobial resistance profile and serotypes of porcine *Salmonella* isolates from Kenyan slaughterhouses, 13.8% were *Salmonella* positive, while 7.1% of isolates tested showed multidrug resistance [72]. Resistance to tetracycline, ampicillin, chloramphenicol and streptomycin were found to be mediated by the *tet*(A), *bla*$_{TEM}$, *catA1* and *strA* genes, respectively [72]. An Ethiopia study recorded a high multidrug resistance value of 36.7% (to seven or more drugs tested) in *Salmonella* isolated from dairy cattle [73]. In a study to determine the prevalence, antimicrobial susceptibility profiles and serotype distribution of faecal *Salmonella* from apparently healthy dogs, Kiflu et al. (2017) found a *Salmonella* carriage rate of 11.7% in dogs screened [74]. Fourteen *Salmonella* serotypes were detected, with the most dominant ones being *S. bronx* (16.7%), and *S. newport* (14.3%) and 9.5% for each of *S. typhimurium*, *S. indiana*, *S. kentucky*, *S. saintpaul* and *S. virchow*. There was an association between *Salmonella* infection and diarrhoeal symptoms in the past 60 days. Highest antibiotic resistance rates were shown against oxytetracycline (59.5%), neomycin (50%) and streptomycin (38.1%), with 45.2% of isolates showing resistance to three or more of the 16 antibiotics tested [74]. Ahmed et al. (2016) detected the virulence genes *stn*, *avrA*, *mgtC*, *invA* and *bcfC* in all screened isolates of *Salmonella enterica* serovar Typhimurium [75]. Antibiotic resistance frequencies detected were as follows; gentamicin (30%), ampicillin and tetracycline (53.3%, each), streptomycin (56.7%) and trimethoprim–sulfamethoxazole and chloramphenicol (73.3%, each). Frequencies of resistance genes discovered in *Salmonella typhimurium*; *sul1* (96.7%), *tet*A(A) (60%), *tet*A(B) (20%), *floR* (73.3%), *aad*A1 (46.7%), *aad*A2 (63.3%), *bla*TEM (53.3%), *aad*B (6.7%) and *aad*C (23.3%) [75].

3.13.2. *Campylobacter*

A study was conducted to determine the antimicrobial resistance profile and epidemiology of *Campylobacter* isolated from humans in Tanzania [76]. The prevalence of *Campylobacter* infection in human samples was 11.4%. A high resistance rate was found against erythromycin (84.3%) and azithromycin (89.6%) whereas a relatively low resistance rate of 22.1% was found against ciprofloxacin [76]. In a Botswana study, phylogenetic analysis showed that *Campylobacter* spp. from different poultry and human sources were highly related [77].

4. Discussion

In Africa, zoonotic diseases remain to be largely neglected by public health and veterinary services, despite causing a substantial health burden in several countries. This work intends to systematically review data on the most important bacterial zoonoses in Africa, within the period of 2008 to 2018, focusing on the presence, prevalence estimates, causative pathogens, control strategies and risk for human infection. We found 58 studies/reports on 29 countries, which were considered of adequate quality to provide estimates of burden of disease or pathogen, with Egypt (eight), Kenya (seven) and Tanzania (six) being the most represented. We found no reports on zoonotic diseases from central African countries eligible to the inclusion criteria. The distribution of bacterial zoonoses studies in the current study was shown in Figure 3 and was found to be in line with previously reported burden of zoonotic diseases in Africa [78]. Although several bacterial zoonoses such as brucellosis, foodborne diseases, Q-fever, and tuberculosis were reported from countries in central Africa [78,79], we found no reports that were eligible to the inclusion criteria on bacterial zoonotic diseases in this region. The current study reviewed data on the evidence of various zoonoses in humans, multiple species of animals, vectors and the environment. Fourteen reports studied possible bacterial zoonoses in humans (including patients visiting hospitals and high-risk groups), 33 reports investigated zoonoses in animals, whereas 11 reports investigated zoonoses in both humans and animals. Nine reports observed the possible roles of vectors in the transmission of bacterial zoonoses. Vector-borne zoonotic bacterial pathogens carried by vectors (ticks, fleas and bat flies) in this study include *Borrelia* spp., *Rickettsia* spp., *C. burnetii*, *Anaplasma* spp. and *Bartonella* spp. The lack of disease surveillance studies and control programs at the national level in most countries introduces a knowledge gap, and makes it difficult to estimate representative disease burden and thoroughly investigate pathogen transmission dynamics. Thus, more national level epidemiological studies ought to be undertaken to bridge this knowledge gap. The epidemiological picture of zoonotic diseases on the African continent is evolving. The prevalence of zoonotic diseases/pathogens summarised in this review must be interpreted with caution, as many of the studies were conducted within specific geographical and occupational settings/groups and may not be extrapolated to the general population. The changing scenes of rapid urbanisations in various countries may translate to the changing epidemiology of zoonotic diseases.

Considering the complex interrelatedness between humans, animals and the environment, any intervention that seeks to tackle the problem of bacterial diseases and antimicrobial resistance from a non-holistic, single focus point of view is bound to fail. The 'One Health' approach seeks to amalgamate and improve the efforts of clinicians, veterinarians, environmentalists, agricultural and public health officials to develop effective surveillance techniques, accompanied by appropriate diagnostic and therapeutic interventions. This holistic and coordinated approach will lead to the enactment of more thorough and effective policies. The achievement of true One-health approach depends of the recognition of the complex interplay between human health, domestic, wild animals, and the environment [78,80–82]. It is crucial to implement the one-health components in low-income and resource-limited countries in Africa to tackle and reduce the increasing threats of bacterial zoonotic infectious diseases [16,83–85]

4.1. Brucellosis

Diagnostic methods most commonly used for brucellosis in developing countries are serologic assays based on rapid slide agglutination tests, albeit the poor specificity of these tests limits their usefulness. Other diagnostic techniques such as ELISA and PCR, were used by most studies on brucellosis [20,36,38,86], are more specific and sensitive, proffering a better correlation with clinical observations, although the latter may not be readily available in many developing countries [87]. The precision of serodiagnosis depends on the presence of antibodies in the serum, and infected animals with low concentrations of antibodies, or no antibodies at all in serum, are therefore likely to present as negative even though they may be infectious [31]. In addition, PCR has the benefit of facilitating the differentiation of *Brucella* genotypes. Considering that the diagnosis and clinical management of febrile illnesses in most developing countries are done empirically, resulting in inaccurate treatment, it is essential to augment the capacity of laboratories to improve the diagnosis accuracy and treatment reliability. This point is highlighted by the fact that 43.2%, 20.5% and 8.2% of patients with brucellosis in the study by Njeru et al. (2016) were diagnosed with typhoid fever, malaria and pneumonia, respectively [20]. In Tanzania, as is the case in many developing countries, brucellosis is an underdiagnosed/misdiagnosed and undertreated disease with no standard treatment protocol usually followed in hospitals, as evinced by the misdiagnosis of it as malaria and pneumonia [21]. The absence of specific symptoms makes it difficult to distinguish brucellosis from several other febrile illnesses occurring in the same geographical area. There is the need for heightened clinical alertness and laboratory capacity building to ensure prompt and accurate diagnoses to aid in the detection and subsequent management of brucellosis in this part of the world. Nasinyama et al. (2014) observed that cELISA test had a sensitivity and specificity of 98.3% and 99.7%, respectively, and is valuable for observing the effectiveness of treatment, prognosis and clinical conditions [36]. Although no single diagnostic test is ideal, with reference to specificity and sensitivity, the standard tube agglutination test (STAT) was preferred in such environments. The limitation of STAT is the long turnaround times, making it unsuitable for seroepidemiological studies, where multiple samples need to be investigated, or in hospital laboratories, where brucellosis therapy has to be initiated quickly. Thus, less time-consuming and faster turnaround diagnostic methods, such as Competitive Enzyme-Linked Assay (cELISA), may need to be used [36].

Although brucellosis has been well recorded in nomadic herdsmen in rural sub-Saharan Africa, owing to their being in close contact with infected animals [88], Bouley et al. (2012) found no association of note between brucellosis and rural residence. While brucellosis prevalence is generally higher in northern Africa [89,90], its seroprevalence ranges from 3 to 8% in sub-Saharan Africa [91]. Despite the implementation of control regimes and strategies, brucellosis remains pervasive in Egypt. Despite immunisations with *Brucella* (*B.*) *abortus* RB51 vaccine, a rise in abortions suspiciously caused by *Brucella* was observed in a dairy cattle herd. The disease has serious economic implications resulting from abortions, infertility and decreased milk production, thus necessitating the implementation of surveillance and control strategies to forestall the socioeconomic effects in both developed and developing countries where the disease is endemic. The prevention, control and eradication strategies against brucellosis usually involve vaccination programmes which employ live, attenuated vaccines as they can elicit long-term cell-mediated immunity [92]. Serological testing and the subsequent culling of seropositive animals are crucial interventions in the adequate control of zoonoses in developing countries.

A large herd size leads to increase in stocking volume, thus exposing more animals to infection [93], as demonstrated by Megersa et al. (2011) [32]. *Brucella* infection in livestock husbandry practice poses zoonotic threats to the public due to close contact with animals, assisting in parturition and the consumption of unpasteurised milk. The study by Osoro et al. (2015) highlights a 'One Health' approach to tackling the menace of brucellosis by concurrently looking into the prevalence of brucellosis in both humans and their livestock in the same household [38]. This approach allows for identification and assessment of risk factors for transmission and gives a more complete epidemiological picture and delineates the factors at play at the human–animal interface [38].

4.2. Q Fever

Q fever is a common cause of febrile illness in Kenya, but it is underestimated [23]. There is a low level of clinical suspicion, with most febrile patients admitted to hospitals given standard empirical treatments that typically include antimalarials and penicillin antibiotics. Even though Njeru et al. (2016) reported a high Q fever prevalence rate of 16.2%, the most common working diagnosis by clinicians documented in this group was typhoid fever (45.1%), followed by acute respiratory infections/pneumonia (37.6%), malaria (6.9%) and fever of unknown origin (10.4%) [23]. There are indications of increasing cases of severe febrile illnesses of under-recognised zoonotic sources facing clinicians, but diagnostic tools for such conditions are lacking in many African countries [94], leading to overdiagnosis of familiar febrile illnesses even when there is no diagnostic evidence to back.

Bok et al. (2017) determined that having at least one seropositive animal (small ruminant) in one's compound was a risk factor for human seropositivity [51], highlighting the relationship between seropositivity and closeness of contact with infected animals. Other studies found risk factors for human infection included exposure to goats, cattle and animal slaughters. Dietary factors linked with seropositivity were found to include consumption of raw milk and locally fermented milk products [23].

The use of point-of-care testing in health care centres will inform treatment and decrease the possibility of wrongful diagnosis and inappropriate treatment in febrile patients seeking treatment at health centres. As shown by Angelakis and colleagues, real-time PCR, which is less time-consuming than conventional PCR, can come in handy in decreasing delays in diagnosis, thereby facilitating prompt treatment [24]. Even though the immunofluorescent assay test (IFAT) is considered the gold standard for serological detection of Q fever, it still falls short and requires highly experienced technicians [52,95]. There is the likelihood that some infected animals may shed bacteria without having antibodies thus they may be classified as negative by serology, leading to an underestimation of associated risk factors. Analysing animals for the shedding ability would partly provide a solution. Excretion of bacteria was found in 60% of flocks by one study [49], presenting a significant risk in the spread of the disease especially to humans.

4.3. Leptospirosis

The possible role of rodents in the transmission of the disease was underscored by the observation that patients with acute infection had a significantly higher contact with rodents than those with presumptive or no infection [26]. Also, a study found an association found between Leptospira infection and rural dwelling (OR 3.4, $p < 0.001$) [27]. Again, a worrying case of misdiagnosis and subsequent inappropriate treatment was observed, as 80% of patients with acute leptospirosis were given antimalarial drugs by prescribers in Mozambique [26].

There may be a gradual expansion in the occurrence of leptospirosis from the typical rural communities to sub-urban communities as evidenced by the fact that 20.9% of the confirmed/presumptive cases of leptospirosis occurred in sub-urban populations in the Mozambique study [26]. This shift has been demonstrated to be associated with inadequate sanitation, poor hygiene, rise in rodent population and poor disposal of solid waste. With the rising trend of rural–urban/sub-urban migration, coupled with attendant problems such as frequent floods and global warming, it can be predicted that leptospirosis will pose a great public health threat in the near future. This prediction is particularly relevant for Mozambique as the country has been rated as the third most vulnerable country to extreme climate events in Africa [26].

MAT as a technique may help provide hints of animal reservoirs by showing the common serogroups prevalent in a specific locality, although the technique is not serovar-specific [96]. In a study by Samir et al. (2015), there was a disagreement between PCR and MAT results in evaluating seroprevalence in humans. This highlights the need for increased surveillance and well-planned prevention and control programs, particularly those that target animals as the source of infection to eradicate the disease. Vaccination programs targeted at livestock and pets would help reduce the disease burden in animals, and reduce environmental contamination and exposure of humans to the pathogen. The detection of *Leptospira interrogans* in banded mongoose (*Mungos mungo*) in Botswana is an important finding, as they are also found frequently in central and eastern Africa, and are thus important to public health [45]. Situations that force humans, domestic animals and wildlife animals to share sources of water put populations at risk of outbreaks, while flooding rivers may carry soil contaminated with urine [45].

4.4. Bovine Tuberculosis

In cattle, post-mortem and bacteriological examinations of suspected lesions are important ways of confirming the presence of bovine tuberculosis. The mycobacterial species concerned are characterised by molecular methods, while the specificity of diagnosis may be improved by histopathological examination. As accurate diagnosis is key, routine culturing and other reliable diagnostic techniques are required to make definitive diagnosis, to help fashion control programs [55]. Phenotype-based characterisation of mycobacteria is laborious and less reproducible compared to molecular detection techniques, such as PCR, which has a higher sensitivity and specificity, and is faster and more reliable [55]. However, conventional detection methods remain useful in many developing countries, as molecular techniques may not be readily available due to cost. It was found that PCR showed high sensitivity and specificity, and thus can be relied upon to confirm the results of tests from Ziehl-Neelsen (ZN) staining, tuberculin skin test and postmortem, particularly as these tests are liable to give false positives [56].

4.5. Rickettsiosis

Ticks and mosquitoes are known to be the two main vectors of several human and animal pathogens [97], with recent studies indicating an increase in the number of tick-borne pathogens of humans and animals [59]. The occurrence of spotted fever group (SFG) rickettsiae differs according to the location and tick gender. The pathogen *Rickettsia felis*, commonly borne by fleas, causes flea-borne spotted fever, which can manifest as a mild to moderate disease, symptoms of which include cutaneous rash, fever, neurologic and digestive signs. Socolovschi et al. (2010) investigated the cause of reported febrile conditions that had tested negative for malaria [61]. The prevalence of spotted fever in all samples was 4.4%, with *R. felis* infection possibly being responsible for many cases of uneruptive fevers of unknown origins particularly those accompanied with digestive, neurologic and respiratory signs [61]. Vector-borne bacterial zoonoses have complex epidemiology and ecology, meaning factors such as weather and climate can affect transmission cycles, making them hard to control [98].

4.6. Anaplasmosis

Ruminants and rodent species are known natural hosts for *Anaplasma phagocytophilum*, with humans and dogs being considered accidental hosts. However, *A. platys* naturally infects dogs, and is thought to be transmitted by the *Rhipicephalus sanguineus* group of ticks [99]. The close bond shared between humans and dogs can facilitate the transmission of pathogens between them, as dogs spend time outdoors and also closely associated with humans, which means that they are a good source of tick-borne infections [64]. The first study to highlight the prevalence of *Anaplasma* spp. in dogs in Zambia [64] is important from the viewpoint of 'One Health', as it recognises dogs as important reservoirs of zoonotic pathogens, thus increasing the risk for human infection. Increased sensitisation among veterinarians and dog owners is essential. Other measures such as use of insect repellents, insecticide treatment of pets and frequent tick checks on pets after outdoor activity in high risk communities will help check the spread of vector-borne pathogens [98].

4.7. Lyme Borreliosis

Lyme borreliosis is mainly transmitted through Ixodes ticks to mammalian hosts. The main reservoirs for the disease are deer and small rodents especially mice. Elhelw et al. (2014) in their study of Lyme borreliosis in Egypt found the *OspA* gene (outer surface protein A gene) and anti-*B. burgdorferi* IgM by PCR and ELISA, respectively, whereas culture identification techniques showed a low sensitivity for the recovery of *Borrelia burgdorferi* isolates in humans [69]. Thus, it would be more tenable to rely on PCR and ELISA when dealing with this pathogen. The prior exposure of human contacts to tick bite in that study, suggests a possible zoonotic transfer.

4.8. Bartonellosis, Plague, Tularaemia and Anthrax

Bartonella species are mostly thought to be transmitted by arthropod vectors. The detection of bacterial DNA, however, does not necessarily indicate that the organism is viable or that the vector is capable of transmitting the pathogen [68].

Plague occurs worldwide, although most suspected human cases are reported in developing countries, with sub-Saharan Africa accounting for more than 95% of the human cases worldwide [29]. In light of the fact that rodents and fleas are natural reservoirs of *Yersinia pestis*—the causative pathogen for plague [29]—Forrester and colleagues found a correlation between reports of human plague and a large number of dead rats in a village, which is unsurprising considering that close contact with infected rodents is a risk factor for the disease. Even though plague is a less frequent zoonosis, it still retains public health significance because of its epidemic potential [98].

As was observed in other studies, most febrile cases that tested positive to tularaemia in a Kenya study [30] were not recognised by clinicians and hence the appropriate treatment protocol was not followed. Indeed, most cases were treated with antimalarials and/or beta-lactam antibiotics which are ineffective against the pathogen of concern.

In developing countries, where there is high level of interaction at the human–animal interface, anthrax, caused by *Bacillus anthracis*, continues to pose public health threats [71]. Testing the susceptibility of bacterial isolates to some antibiotics, Hang'ombe et al. (2012) in an investigation of a suspected anthrax outbreak, observed that all tested isolates were sensitive to the antibiotics used (including ciprofloxacin and doxycycline), except vancomycin. Ciprofloxacin and doxycycline are recommended by the US Centers for Disease Control and Prevention (CDC) as first line treatment for anthrax [100].

4.9. Other Zoonotic Pathogens

Other bacterial zoonotic pathogens, including *Salmonella* spp. and *Campylobacter* spp., which can be transmitted between livestock and humans, were reported by various studies.

Salmonellosis is one of the most common foodborne zoonoses in developing and industrialised countries [72]. The presence of *Salmonella* in food animals and animal products presents a food safety threat [72]. Food safety measures need to be intensified particularly as multidrug resistant pathogenic strains are increasing.

Campylobacter frequently colonizes different species of animals asymptomatically, but produces acute and self-limiting intestinal infections in humans [76], with undercooked and raw poultry meat having been particularly found to be culpable. In a study by Komba et al. (2015), *Campylobacter* isolates showed 84.3% resistance to erythromycin, which is worrying, considering that erythromycin together with ciprofloxacin are the antibiotics of choice in the treatment of severe, nonself-limiting *Campylobacter* infections such as septic arthritis, bacteremia and prolonged enteritis [76,101]. Salmonellosis and campylobacteriosis are reported as the commonest foodborne bacterial zoonoses in Europe with eggs and mixed foods as the most culpable food sources [98]. However, prevalence data for these two zoonoses are lacking in Africa. The overuse of antibiotics (mainly as growth promoters) in animal husbandry, coupled with the close contact of humans and farm animals, facilitates the emergence of resistant zoonotic bacterial pathogens. Indeed, studies elsewhere have shown that resistance in pathogenic zoonotic bacteria and/or changes in faecal microbiota increases shortly after the introduction of antibiotics in veterinary practice [102–104]. Stricter controls concerning the nontherapeutic use of antibiotics in animal husbandry are required.

4.10. Limitations of the Data

The lack of surveys on zoonoses at the national levels, as well as individual studies not being representative enough, might affect the true estimates of zoonoses in individual countries and across the continent. Furthermore, individual reports included in this study have not factored in confounding bias, which may affect the true estimates.

5. Conclusions

Bacterial zoonotic diseases pose a significant burden in Africa, although the actual socioeconomic burden is unknown. Interactions at the human–livestock and human–wildlife interfaces contribute to the transmission of zoonoses, with a wide range of hosts and vectors playing roles. Bacterial zoonoses have a dual impact on both livestock production systems and human health. The lack of diagnostic tests and clinical awareness for many zoonotic diseases in most parts of Africa is worrying, being reflected in the low levels of diagnoses on the continent in clinical settings. A 'One Health' approach, which involves the concerted efforts of veterinarians, physicians, public health workers and epidemiologists, is essential in the policy schemes that are aimed at controlling and preventing the transmission of such diseases.

Author Contributions: Conceptualization, M.E.E.Z.; methodology, J.A. and M.E.E.Z.; validation, J.A., A.N. and M.E.E.Z.; formal analysis, J.A. and M.E.Z; investigation, J.A. and M.E.E.Z.; resources, M.E.E.Z. and A.N.; data curation, J.A. and M.E.E.Z.; writing—original draft preparation, J.A. and M.E.E.Z.; writing—review and editing, M.E.E.Z., J.A. and A.N.; visualization, J.A. and M.E.E.Z.; supervision, M.E.E.Z.; project administration, M.E.E.Z.

Funding: This research received no external funding.

Acknowledgments: The authors would like to thank the two anonymous reviewers for their comments that improved the manuscript. Authors would like to thank Zelalem G. Dessie from the School of Mathematics, Statistics and Computer Science for his assistance.

Conflicts of Interest: The authors declare no conflict of interest.

References

1. Simpson, G.J.; Quan, V.; Frean, J.; Knobel, D.L.; Rossouw, J.; Weyer, J.; Marcotty, T.; Godfroid, J.; Blumberg, L.H. Prevalence of selected zoonotic diseases and risk factors at a human-wildlife-livestock interface in Mpumalanga Province, South Africa. *Vector-Borne Zoonotic Dis.* **2018**, *18*, 303–310. [CrossRef]
2. Jones, K.E.; Patel, N.G.; Levy, M.A.; Storeygard, A.; Balk, D.; Gittleman, J.L.; Daszak, P. Global trends in emerging infectious diseases. *Nature* **2008**, *451*, 990–993. [CrossRef] [PubMed]
3. Grace, D.; Gilbert, J.; Randolph, T.; Kang'ethe, E. The multiple burdens of zoonotic disease and an ecohealth approach to their assessment. *Trop. Anim. Health Prod.* **2012**, *44*, 67–73. [CrossRef]
4. Grace, D.; Mutua, F.; Ochungo, P.; Kruska, R.; Jones, K.; Brierley, L.; Lapar, M.; Said, M.; Herrero, M.; Phuc, P. *Mapping of Poverty and Likely Zoonoses Hotspots*; ILRI: Nairobi, Kenya, 2012.
5. Grace, D.; Lindahl, J.; Wanyoike, F.; Bett, B.; Randolph, T.; Rich, K.M. Poor livestock keepers: Ecosystem–poverty–health interactions. *Phil. Trans. R. Soc. B* **2017**, *372*, 20160166. [CrossRef]
6. WHO. *Who Estimates of the Global Burden of Foodborne Diseases: Foodborne Disease Burden Epidemiology Reference Group 2007–2015*; WHO: Geneva, Switzerland, 2015.
7. Fèvre, E.M.; Glanville, W.A.; Thomas, L.F.; Cook, E.A.; Kariuki, S.; Wamae, C.N. An integrated study of human and animal infectious disease in the Lake Victoria crescent small-holder crop-livestock production system, Kenya. *BMC Infect. Dis.* **2017**, *17*, 457. [CrossRef] [PubMed]

8. Munyua, P.; Bitek, A.; Osoro, E.; Pieracci, E.G.; Muema, J.; Mwatondo, A.; Kungu, M.; Nanyingi, M.; Gharpure, R.; Njenga, K. Prioritization of zoonotic diseases in Kenya, 2015. *PLoS ONE* **2016**, *11*, e0161576. [CrossRef]

9. Centers for Disease Control and Prevention. Bioterrorism Agents/Diseases. Emergency Preparedness and Response. Centers for Disease Control and Prevention: Atlanta, GA, USA. Available online: https://emergency.cdc.gov/agent/agentlist-category.asp (accessed on 9 April 2018).

10. Rodríguez-Prieto, V.; Vicente-Rubiano, M.; Sánchez-Matamoros, A.; Rubio-Guerri, C.; Melero, M.; Martínez-López, B.; Martínez-Avilés, M.; Hoinville, L.; Vergne, T.; Comin, A. Systematic review of surveillance systems and methods for early detection of exotic, new and re-emerging diseases in animal populations. *Epidemiol. Infect.* **2015**, *143*, 2018–2042. [CrossRef]

11. Zumla, A.; Dar, O.; Kock, R.; Muturi, M.; Ntoumi, F.; Kaleebu, P.; Eusebio, M.; Mfinanga, S.; Bates, M.; Mwaba, P. *Taking Forward a 'One Health' approach for Turning the Tide Against the Middle East Respiratory Syndrome Coronavirus and Other Zoonotic Pathogens with Epidemic Potential*; Elsevier: Amsterdam, The Netherlands, 2016.

12. Dweba, C.C.; Zishiri, O.T.; El Zowalaty, M.E. Methicillin-resistant *Staphylococcus aureus*: Livestock-associated, antimicrobial, and heavy metal resistance. *Infect. Drug Resist.* **2018**, *11*, 2479.

13. Holmes, A.H.; Moore, L.S.; Sundsfjord, A.; Steinbakk, M.; Regmi, S.; Karkey, A.; Guerin, P.J.; Piddock, L.J. Understanding the mechanisms and drivers of antimicrobial resistance. *Lancet* **2016**, *387*, 176–187. [CrossRef]

14. Laxminarayan, R.; Duse, A.; Wattal, C.; Zaidi, A.K.; Wertheim, H.F.; Sumpradit, N.; Vlieghe, E.; Hara, G.L.; Gould, I.M.; Goossens, H. Antibiotic resistance—The need for global solutions. *Lancet Infect. Dis.* **2013**, *13*, 1057–1098. [CrossRef]

15. Vanderburg, S.; Rubach, M.P.; Halliday, J.E.; Cleaveland, S.; Reddy, E.A.; Crump, J.A. Epidemiology of *Coxiella burnetii* infection in Africa: A one health systematic review. *PLoS Negl. Trop. Dis.* **2014**, *8*, e2787. [CrossRef]

16. Cleaveland, S.; Sharp, J.; Abela-Ridder, B.; Allan, K.J.; Buza, J.; Crump, J.A.; Davis, A.; Del Rio Vilas, V.J.; de Glanville, W.A.; Kazwala, R.R.; et al. One Health contributions towards more effective and equitable approaches to health in low-and middle-income countries. *Philos. Trans. R. Soc. B: Biol. Sci.* **2017**, *372*, 20160168.

17. United Nations Statistics Division. *Composition of Macro Geographical (Continental) Regions, Geographical Sub-Regions, and Selected Economic and Other Groupings*; United Nations Statistics Division: New York, NY, USA, 2010.

18. IBM Corp. Released 2017. *IBM SPSS Statistics for Windows*, Version 25.0; IBM Corp: Armonk, NY, USA, 2017.

19. R Development Core Team. *R: A Language and Environment for Statistical Computing*; R Foundation for Statistical Computing: Vienna, Austria, 2008; ISBN 3-900051-07-0.

20. Njeru, J.; Melzer, F.; Wareth, G.; El-Adawy, H.; Henning, K.; Pletz, M.W.; Heller, R.; Kariuki, S.; Fèvre, E.; Neubauer, H. Human brucellosis in febrile patients seeking treatment at remote hospitals, Northeastern Kenya, 2014–2015. *Emerg. Infect. Dis.* **2016**, *22*, 2160–2164. [CrossRef]

21. Bouley, A.J.; Biggs, H.M.; Stoddard, R.A.; Morrissey, A.B.; Bartlett, J.A.; Afwamba, I.A.; Maro, V.P.; Kinabo, G.D.; Saganda, W.; Cleaveland, S. Brucellosis among hospitalized febrile patients in Northern Tanzania. *Am. J. Trop. Med. Hyg.* **2012**, *87*, 1105–1111. [CrossRef]

22. Chipwaza, B.; Mhamphi, G.G.; Ngatunga, S.D.; Selemani, M.; Amuri, M.; Mugasa, J.P.; Gwakisa, P.S. Prevalence of bacterial febrile illnesses in children in Kilosa District, Tanzania. *PLoS Negl. Trop. Dis.* **2015**, *9*, e0003750. [CrossRef]

23. Njeru, J.; Henning, K.; Pletz, M.; Heller, R.; Forstner, C.; Kariuki, S.; Fèvre, E.; Neubauer, H. Febrile patients admitted to remote hospitals in northeastern Kenya: Seroprevalence, risk factors and a clinical prediction tool for Q-fever. *BMC Infect. Dis.* **2016**, *16*, 244. [CrossRef]

24. Angelakis, E.; Mediannikov, O.; Socolovschi, C.; Mouffok, N.; Bassene, H.; Tall, A.; Niangaly, H.; Doumbo, O.; Znazen, A.; Sarih, M. *Coxiella burnetii*-positive PCR in febrile patients in rural and urban Africa. *Int. J. Infect. Dis.* **2014**, *28*, 107–110. [CrossRef]

25. Prabhu, M.; Nicholson, W.L.; Roche, A.J.; Kersh, G.J.; Fitzpatrick, K.A.; Oliver, L.D.; Massung, R.F.; Morrissey, A.B.; Bartlett, J.A.; Onyango, J.J. Q fever, spotted fever group, and typhus group rickettsioses among hospitalized febrile patients in northern Tanzania. *Clin. Infect. Dis.* **2011**, *53*, e8–e15. [CrossRef]

26. Ribeiro, P.; Bhatt, N.; Ali, S.; Monteiro, V.; da Silva, E.; Balassiano, I.T.; Aquino, C.; de Deus, N.; Guiliche, O.; Muianga, A.F. Seroepidemiology of leptospirosis among febrile patients in a rapidly growing sub-urban slum and a flood-vulnerable rural district in Mozambique, 2012–2014: Implications for the management of fever. *Int. J. Infect. Dis.* **2017**, *64*, 50–57. [CrossRef]

27. Biggs, H.M.; Bui, D.M.; Galloway, R.L.; Stoddard, R.A.; Shadomy, S.V.; Morrissey, A.B.; Bartlett, J.A.; Onyango, J.J.; Maro, V.P.; Kinabo, G.D. Leptospirosis among hospitalized febrile patients in Northern Tanzania. *Am. J. Trop. Med. Hyg.* **2011**, *85*, 275–281. [CrossRef]

28. Samir, A.; Soliman, R.; El-Hariri, M.; Abdel-Moein, K.; Hatem, M.E. Leptospirosis in animals and human contacts in Egypt: Broad range surveillance. *Rev. Soc. Bras. Med. Trop.* **2015**, *48*, 272–277. [CrossRef]

29. Forrester, J.D.; Apangu, T.; Griffith, K.; Acayo, S.; Yockey, B.; Kaggwa, J.; Kugeler, K.J.; Schriefer, M.; Sexton, C.; Beard, C.B. Patterns of human plague in Uganda, 2008–2016. *Emerg. Infect. Dis.* **2017**, *23*, 1517–1521. [CrossRef]

30. Njeru, J.; Tomaso, H.; Mertens, K.; Henning, K.; Wareth, G.; Heller, R.; Kariuki, S.; Fèvre, E.M.; Neubauer, H.; Pletz, M. Serological evidence of *Francisella tularensis* in febrile patients seeking treatment at remote hospitals, northeastern Kenya, 2014–2015. *New Microbes New Infect.* **2017**, *19*, 62–66. [CrossRef]

31. Wareth, G.; Melzer, F.; Böttcher, D.; El-Diasty, M.; El-Beskawy, M.; Rasheed, N.; Schmoock, G.; Roesler, U.; Sprague, L.D.; Neubauer, H. Molecular typing of isolates obtained from aborted foetuses in *Brucella*-free Holstein dairy cattle herd after immunisation with *Brucella abortus* RB51 vaccine in Egypt. *Acta Trop.* **2016**, *164*, 267–271. [CrossRef]

32. Megersa, B.; Biffa, D.; Niguse, F.; Rufael, T.; Asmare, K.; Skjerve, E. Cattle brucellosis in traditional livestock husbandry practice in southern and eastern Ethiopia, and its zoonotic implication. *Acta Vet. Scand.* **2011**, *53*, 24. [CrossRef]

33. Wareth, G.; Melzer, F.; Elschner, M.C.; Neubauer, H.; Roesler, U. Detection of *Brucella melitensis* in bovine milk and milk products from apparently healthy animals in Egypt by real-time PCR. *J. Infect. Dev. Ctries* **2014**, *8*, 1339–1343. [CrossRef]

34. Erume, J.; Roesel, K.; Dione, M.M.; Ejobi, F.; Mboowa, G.; Kungu, J.M.; Akol, J.; Pezo, D.; El-Adawy, H.; Melzer, F. Serological and molecular investigation for brucellosis in swine in selected districts of Uganda. *Trop. Anim. Health Prod.* **2016**, *48*, 1147–1155. [CrossRef]

35. Wareth, G.; Melzer, F.; Tomaso, H.; Roesler, U.; Neubauer, H. Detection of *Brucella abortus* DNA in aborted goats and sheep in Egypt by real-time PCR. *BMC Res. Notes* **2015**, *8*, 212. [CrossRef]

36. Nasinyama, G.; Ssekawojwa, E.; Opuda, J.; Grimaud, P.; Etter, E.; Bellinguez, A. *Brucella* sero-prevalence and modifiable risk factors among predisposed cattle keepers and consumers of un-pasteurized milk in Mbarara and Kampala Districts, Uganda. *Afr. Health Sci.* **2014**, *14*, 790–796. [CrossRef]

37. Sylla, S.; Sidimé, Y.; Sun, Y.; Doumbouya, S.; Cong, Y. Seroprevalence investigation of bovine brucellosis in Macenta and Yomou, Guinea. *Trop. Anim. Health Prod.* **2014**, *46*, 1185–1191. [CrossRef]

38. Osoro, E.M.; Munyua, P.; Omulo, S.; Ogola, E.; Ade, F.; Mbatha, P.; Mbabu, M.; Kairu, S.; Maritim, M.; Thumbi, S.M. Strong association between human and animal *Brucella* seropositivity in a linked study in Kenya, 2012–2013. *Am. J. Trop. Med. Hyg.* **2015**, *93*, 224–231. [CrossRef]

39. Kanouté, Y.B.; Gragnon, B.G.; Schindler, C.; Bonfoh, B.; Schelling, E. Epidemiology of brucellosis, Q fever and Rift Valley fever at the human and livestock interface in northern Côte d'Ivoire. *Acta Trop.* **2017**, *165*, 66–75. [CrossRef]

40. Boone, I.; Henning, K.; Hilbert, A.; Neubauer, H.; Von Kalckreuth, V.; Dekker, D.M.; Schwarz, N.G.; Pak, G.D.; Krüger, A.; Hagen, R.M. Are brucellosis, Q fever and melioidosis potential causes of febrile illness in Madagascar? *Acta Trop.* **2017**, *172*, 255–262. [CrossRef]

41. Dean, A.S.; Bonfoh, B.; Kulo, A.E.; Boukaya, G.A.; Amidou, M.; Hattendorf, J.; Pilo, P.; Schelling, E. Epidemiology of brucellosis and Q fever in linked human and animal populations in northern Togo. *PLoS ONE* **2013**, *8*, e71501. [CrossRef]

42. Tebug, S.; Njunga, G.R.; Chagunda, M.G.; Mapemba, J.P.; Awah-Ndukum, J.; Wiedemann, S. Risk, knowledge and preventive measures of smallholder dairy farmers in northern Malawi with regard to zoonotic brucellosis and bovine tuberculosis. *Onderstepoort J. Vet. Res.* **2014**, *81*, 1–6. [CrossRef]

43. Lagadec, E.; Gomard, Y.; Guernier, V.; Dietrich, M.; Pascalis, H.; Temmam, S.; Ramasindrazana, B.; Goodman, S.M.; Tortosa, P.; Dellagi, K. Pathogenic *Leptospira* spp. in bats, Madagascar and Union of the Comoros. *Emerg. Infect. Dis.* **2012**, *18*, 1696–1698. [CrossRef]

44. Dietrich, M.; Wilkinson, D.A.; Soarimalala, V.; Goodman, S.M.; Dellagi, K.; Tortosa, P. Diversification of an emerging pathogen in a biodiversity hotspot: *L. eptospira* in endemic small mammals of Madagascar. *Mol. Ecol.* **2014**, *23*, 2783–2796. [CrossRef]

45. Jobbins, S.; Sanderson, C.; Alexander, K. *Leptospira interrogans* at the human–wildlife interface in northern Botswana: A newly identified public health threat. *Zoonoses Public Health* **2014**, *61*, 113–123. [CrossRef]

46. Dobigny, G.; Garba, M.; Tatard, C.; Loiseau, A.; Galan, M.; Kadaouré, I.; Rossi, J.-P.; Picardeau, M.; Bertherat, E. Urban market gardening and rodent-borne pathogenic *Leptospira* in arid zones: A case study in Niamey, Niger. *PLoS Negl. Trop. Dis.* **2015**, *9*, e0004097. [CrossRef]

47. Allan, K.J.; Halliday, J.E.; Moseley, M.; Carter, R.W.; Ahmed, A.; Goris, M.G.; Hartskeerl, R.A.; Keyyu, J.; Kibona, T.; Maro, V.P. Assessment of animal hosts of pathogenic Leptospira in northern Tanzania. *PLoS Negl. Trop. Dis.* **2018**, *12*, e0006444. [CrossRef]

48. Abdel-Moein, K.A.; Hamza, D.A. The burden of *Coxiella burnetii* among aborted dairy animals in Egypt and its public health implications. *Acta Trop.* **2017**, *166*, 92–95. [CrossRef]

49. Khaled, H.; Sidi-Boumedine, K.; Merdja, S.; Dufour, P.; Dahmani, A.; Thiéry, R.; Rousset, E.; Bouyoucef, A. Serological and molecular evidence of Q fever among small ruminant flocks in Algeria. *Comp. Immunol. Microbiol. Infect. Dis.* **2016**, *47*, 19–25. [CrossRef]

50. Abushahba, M.F.; Abdelbaset, A.E.; Rawy, M.S.; Ahmed, S.O. Cross-sectional study for determining the prevalence of Q fever in small ruminants and humans at El Minya Governorate, Egypt. *BMC Res. Notes* **2017**, *10*, 538. [CrossRef] [PubMed]

51. Bok, J.; Hogerwerf, L.; Germeraad, E.A.; Roest, H.I.; Faye-Joof, T.; Jeng, M.; Nwakanma, D.; Secka, A.; Stegeman, A.; Goossens, B. *Coxiella burnetii* (Q fever) prevalence in associated populations of humans and small ruminants in the Gambia. *Trop. Med. Int. Health* **2017**, *22*, 323–331. [CrossRef] [PubMed]

52. Wardrop, N.A.; Thomas, L.F.; Cook, E.A.; de Glanville, W.A.; Atkinson, P.M.; Wamae, C.N.; Fèvre, E.M. The sero-epidemiology of *Coxiella burnetii* in humans and cattle, western Kenya: Evidence from a cross-sectional study. *Plos Negl. Trop. Dis.* **2016**, *10*, e0005032. [CrossRef]

53. Mtshali, K.; Nakao, R.; Sugimoto, C.; Thekisoe, O. Occurrence of *Coxiella burnetii*, *Ehrlichia canis*, *Rickettsia* species and *Anaplasma phagocytophilum*-like bacterium in ticks collected from dogs and cats in South Africa. *J. S. Afr. Vet. Assoc.* **2017**, *88*, 1–6. [CrossRef] [PubMed]

54. Kamani, J.; Baneth, G.; Gutiérrez, R.; Nachum-Biala, Y.; Mumcuoglu, K.Y.; Harrus, S. *Coxiella burnetii* and *Rickettsia conorii*: Two zoonotic pathogens in peridomestic rodents and their ectoparasites in Nigeria. *Ticks Tick-Borne Dis.* **2018**, *9*, 86–92. [CrossRef] [PubMed]

55. El Tigani, A.; El Sanousi, S.M.; Gameel, A.; El Beir, H.; Fathelrahman, M.; Terab, N.M.; Muaz, M.A.; Hamid, M.E. Bovine tuberculosis in South Darfur State, Sudan: An abattoir study based on microscopy and molecular detection methods. *Trop. Anim. Health Prod.* **2013**, *45*, 469–472.

56. Sa'idu, A.; Okolocha, E.; Dzikwi, A.; Kwaga, J.; Gamawa, A.; Usman, A.; Maigari, S.; Ibrahim, S. Detection of *Mycobacterium bovis* in organs of slaughtered cattle by DNA-based polymerase chain reaction and Ziehl-Neelsen techniques in Bauchi State, Nigeria. *J. Vet. Med.* **2015**, *2015*. [CrossRef]

57. Okeke, L.A.; Fawole, O.; Muhammad, M.; Okeke, I.O.; Nguku, P.; Wasswa, P.; Dairo, D.; Cadmus, S. Bovine tuberculosis: A retrospective study at Jos abattoir, Plateau State, Nigeria. *Pan Afr. Med J.* **2016**, *25*. [CrossRef]

58. Okeke, L.A.; Cadmus, S.; Okeke, I.O.; Muhammad, M.; Awoloh, O.; Dairo, D.; Waziri, E.N.; Olayinka, A.; Nguku, P.M.; Fawole, O. Prevalence and risk factors of *Mycobacterium tuberculosis* complex infection in slaughtered cattle at Jos South abattoir, Plateau State, Nigeria. *Pan Afr. Med. J.* **2014**, *18*. [CrossRef]

59. Kumsa, B.; Socolovschi, C.; Raoult, D.; Parola, P. Spotted fever group Rickettsiae in Ixodid ticks in Oromia, Ethiopia. *Ticks Tick-Borne Dis.* **2015**, *6*, 8–15. [CrossRef]

60. Moumouni, P.F.A.; Terkawi, M.A.; Jirapattharasate, C.; Cao, S.; Liu, M.; Nakao, R.; Umemiya-Shirafuji, R.; Yokoyama, N.; Sugimoto, C.; Fujisaki, K. Molecular detection of spotted fever group rickettsiae in *Amblyomma variegatum* ticks from Benin. *Ticks Tick-Borne Dis.* **2016**, *7*, 828–833. [CrossRef] [PubMed]

61. Socolovschi, C.; Mediannikov, O.; Sokhna, C.; Tall, A.; Diatta, G.; Bassene, H.; Trape, J.-F.; Raoult, D. *Rickettsia felis*—Associated uneruptive fever, Senegal. *Emerg. Infect. Dis.* **2010**, *16*, 1140. [CrossRef] [PubMed]

62. Mwamuye, M.M.; Kariuki, E.; Omondi, D.; Kabii, J.; Odongo, D.; Masiga, D.; Villinger, J. Novel rickettsia and emergent tick-borne pathogens: A molecular survey of ticks and tick-borne pathogens in Shimba Hills National Reserve, Kenya. *Ticks Tick-Borne Dis.* **2017**, *8*, 208–218. [CrossRef] [PubMed]

63. Nakayima, J.; Hayashida, K.; Nakao, R.; Ishii, A.; Ogawa, H.; Nakamura, I.; Moonga, L.; Hang'ombe, B.M.; Mweene, A.S.; Thomas, Y. Detection and characterization of zoonotic pathogens of free-ranging non-human primates from Zambia. *Parasites Vectors* **2014**, *7*, 490. [CrossRef] [PubMed]

64. Vlahakis, P.A.; Chitanga, S.; Simuunza, M.C.; Simulundu, E.; Qiu, Y.; Changula, K.; Chambaro, H.M.; Kajihara, M.; Nakao, R.; Takada, A. Molecular detection and characterization of zoonotic *Anaplasma* species in domestic dogs in Lusaka, Zambia. *Ticks Tick-Borne Dis.* **2018**, *9*, 39–43. [CrossRef]

65. Said, M.B.; Belkahia, H.; El Mabrouk, N.; Saidani, M.; Alberti, A.; Zobba, R.; Cherif, A.; Mahjoub, T.; Bouattour, A.; Messadi, L. *Anaplasma platys*-like strains in ruminants from Tunisia. *Infect. Genet. Evol.* **2017**, *49*, 226–233. [CrossRef]

66. Kelly, P.; Marabini, L.; Dutlow, K.; Zhang, J.; Loftis, A.; Wang, C. Molecular detection of tick-borne pathogens in captive wild felids, Zimbabwe. *Parasites Vectors* **2014**, *7*, 514. [CrossRef]

67. Kamani, J.; Baneth, G.; Mitchell, M.; Mumcuoglu, K.Y.; Gutiérrez, R.; Harrus, S. Bartonella species in bats (Chiroptera) and bat flies (Nycteribiidae) from Nigeria, West Africa. *Vector-Borne Zoonotic Dis.* **2014**, *14*, 625–632. [CrossRef]

68. Billeter, S.; Hayman, D.; Peel, A.; Baker, K.; Wood, J.; Cunningham, A.; Suu-Ire, R.; Dittmar, K.; Kosoy, M. *Bartonella* species in bat flies (Diptera: Nycteribiidae) from Western Africa. *Parasitology* **2012**, *139*, 324–329. [CrossRef]

69. Elhelw, R.A.; El-Enbaawy, M.I.; Samir, A. Lyme borreliosis: A neglected zoonosis in Egypt. *Acta Trop.* **2014**, *140*, 188–192. [CrossRef]

70. Schwan, T.G.; Anderson, J.M.; Lopez, J.E.; Fischer, R.J.; Raffel, S.J.; McCoy, B.N.; Safronetz, D.; Sogoba, N.; Maïga, O.; Traoré, S.F. Endemic foci of the tick-borne relapsing fever spirochete *Borrelia crocidurae* in Mali, West Africa, and the potential for human infection. *Plos Negl. Trop. Dis.* **2012**, *6*, e1924. [CrossRef]

71. Hang'ombe, M.B.; Mwansa, J.C.; Muwowo, S.; Mulenga, P.; Kapina, M.; Musenga, E.; Squarre, D.; Mataa, L.; Thomas, S.Y.; Ogawa, H. Human–animal anthrax outbreak in the Luangwa Valley of Zambia in 2011. *Trop. Doct.* **2012**, *42*, 136–139. [CrossRef]

72. Ombui, J.N.; Mitema, E.S.; Kikuvi, G.M. Serotypes and antimicrobial resistance profiles of *Salmonella* isolates from pigs at slaughter in Kenya. *J. Infect. Dev. Ctries* **2010**, *4*, 243–248.

73. Eguale, T.; Engidawork, E.; Gebreyes, W.A.; Asrat, D.; Alemayehu, H.; Medhin, G.; Johnson, R.P.; Gunn, J.S. Fecal prevalence, serotype distribution and antimicrobial resistance of *Salmonellae* in dairy cattle in central Ethiopia. *BMC Microbiol.* **2016**, *16*, 20. [CrossRef]

74. Kiflu, B.; Alemayehu, H.; Abdurahaman, M.; Negash, Y.; Eguale, T. *Salmonella* serotypes and their antimicrobial susceptibility in apparently healthy dogs in Addis Ababa, Ethiopia. *BMC Vet. Res.* **2017**, *13*, 134. [CrossRef]

75. Ahmed, H.A.; El-Hofy, F.I.; Shafik, S.M.; Abdelrahman, M.A.; Elsaid, G.A. Characterization of virulence-associated genes, antimicrobial resistance genes, and class 1 integrons in *Salmonella enterica* serovar typhimurium isolates from chicken meat and humans in Egypt. *Foodborne Pathog. Dis.* **2016**, *13*, 281–288. [CrossRef]

76. Komba, E.V.; Mdegela, R.H.; Msoffe, P.; Nielsen, L.N.; Ingmer, H. Prevalence, antimicrobial resistance and risk factors for thermophilic *Campylobacter* infections in symptomatic and asymptomatic humans in Tanzania. *Zoonoses Public Health* **2015**, *62*, 557–568. [CrossRef]

77. De Vries, S.P.; Vurayai, M.; Holmes, M.; Gupta, S.; Bateman, M.; Goldfarb, D.; Maskell, D.J.; Matsheka, M.I.; Grant, A.J. Phylogenetic analyses and antimicrobial resistance profiles of *Campylobacter* spp. from diarrhoeal patients and chickens in Botswana. *PLoS ONE* **2018**, *13*, e0194481. [CrossRef]

78. Grace, D.; Mutua, F.; Ochungo, P.; Kruska, R.L.; Jones, K.; Brierley, L.; Lapar, M.; Said, M.Y.; Herrero, M.T.; Phuc, P.M.; et al. Mapping of poverty and likely zoonoses hotspots. Project 4. Report to the UK Department for International Development. International Livestock Research Institute: Nairobi, Kenya, 2012. Available online: http://cgspace.cgiar.org/handle/10568/21161 (accessed on 11 April 2019).

79. Kelly, A.; Osburn, B.; Salman, M. Veterinary medicine's increasing role in global health. *Lancet Glob. Health.* **2014**, *2*, e379–e380. [CrossRef]

80. Gebreyes, W.A.; Dupouy-Camet, J.; Newport, M.J.; Oliveira, C.J.; Schlesinger, L.S.; Saif, Y.M.; Kariuki, S.; Saif, L.J.; Saville, W.; Wittum, T.; et al. The global one health paradigm: Challenges and opportunities for tackling infectious diseases at the human, animal, and environment interface in low-resource settings. *Plos Negl. Trop. Dis.* **2014**, *8*, e3257. [CrossRef]

81. Belay, E.D.; Kile, J.C.; Hall, A.J.; Barton-Behravesh, C.; Parsons, M.B.; Salyer, S.; Walke, H. Zoonotic disease programs for enhancing global health security. *Emerg. Infect. Dis.* **2017**, *23*, S65.

82. Cunningham, A.A.; Daszak, P.; Wood, J.L. One Health, emerging infectious diseases and wildlife: Two decades of progress? *Philos. Trans. R. Soc. B: Biol. Sci.* **2017**, *372*, 20160167. [CrossRef]

83. World Bank. *People, Pathogens and Our Planet. Volume 2, The Economics of One Health. World Bank Report (#69145-GLB)*; World Bank: Washington, DC, USA, 2012; Available online: http://documents.worldbank.org/curated/en/612341468147856529/People-pathogens-and-our-planet-the-economics-of-one-health (accessed on 11 April 2019).

84. Coker, R.; Rushton, J.; Mounier-Jack, S.; Karimuribo, E.; Lutumba, P.; Kambarage, D.; Pfeiffer, D.U.; Stärk, K.; Rweyemamu, M. Towards a conceptual framework to support one-health research for policy on emerging zoonoses. *Lancet Infect. Dis.* **2011**, *11*, 326–331. [CrossRef]

85. Molyneux, D.; Hallaj, Z.; Keusch, G.T.; McManus, D.P.; Ngowi, H.; Cleaveland, S.; Ramos-Jimenez, P.; Gotuzzo, E.; Kar, K.; Sanchez, A.; et al. Zoonoses and marginalised infectious diseases of poverty: Where do we stand? *Parasites Vectors* **2011**, *4*, 106.

86. Wareth, G.; Hikal, A.; Refai, M.; Melzer, F.; Roesler, U.; Neubauer, H. Animal brucellosis in Egypt. *J. Infect. Dev. Ctries* **2014**, *8*, 1365–1373. [CrossRef]

87. Al, S.D.; Tomaso, H.; Nöckler, K.; Neubauer, H.; Frangoulidis, D. Laboratory-based diagnosis of brucellosis—A review of the literature. Part II: Serological tests for brucellosis. *Clin. Lab.* **2003**, *49*, 577–589.

88. Ari, M.D.; Guracha, A.; Fadeel, M.A.; Njuguna, C.; Njenga, M.K.; Kalani, R.; Abdi, H.; Warfu, O.; Omballa, V.; Tetteh, C. Challenges of establishing the correct diagnosis of outbreaks of acute febrile illnesses in Africa: The case of a likely *Brucella* outbreak among nomadic pastoralists, Northeast Kenya, March–July 2005. *Am. J. Trop. Med. Hyg.* **2011**, *85*, 909–912. [CrossRef]

89. Reddy, E.A.; Shaw, A.V.; Crump, J.A. Community-acquired bloodstream infections in Africa: A systematic review and meta-analysis. *Lancet Infect. Dis.* **2010**, *10*, 417–432. [CrossRef]

90. John, K.; Fitzpatrick, J.; French, N.; Kazwala, R.; Kambarage, D.; Mfinanga, G.S.; MacMillan, A.; Cleaveland, S. Quantifying risk factors for human brucellosis in rural northern Tanzania. *PLoS ONE* **2010**, *5*, e9968. [CrossRef]

91. Swai, E.S.; Schoonman, L. Human brucellosis: Seroprevalence and risk factors related to high risk occupational groups in Tanga Municipality, Tanzania. *Zoonoses Public Health* **2009**, *56*, 183–187. [CrossRef] [PubMed]

92. Siadat, S.D.; Salmani, A.S.; Aghasadeghi, M.R. Brucellosis vaccines: An overview. In *Zoonosis*; Intech: Rijeka, Croatia, 2012.

93. Berhe, G.; Belihu, K.; Asfaw, Y. Seroepidemiological investigation of bovine brucellosis in the extensive cattle production system of Tigray region of Ethiopia. *Int. J. Appl. Res. Vet. Med.* **2007**, *5*, 65.

94. Onchiri, F.M.; Pavlinac, P.B.; Singa, B.O.; Naulikha, J.M.; Odundo, E.A.; Farquhar, C.; Richardson, B.A.; John-Stewart, G.; Walson, J.L. Frequency and correlates of malaria over-treatment in areas of differing malaria transmission: A cross-sectional study in rural western Kenya. *Malar. J.* **2015**, *14*, 97. [CrossRef] [PubMed]

95. Raoult, D. Chronic Q fever: Expert opinion versus literature analysis and consensus. *J. Infect.* **2012**, *65*, 102–108. [CrossRef]

96. Smythe, L.D.; Wuthiekanun, V.; Chierakul, W.; Suputtamongkol, Y.; Tiengrim, S.; Dohnt, M.F.; Symonds, M.L.; Slack, A.T.; Apiwattanaporn, A.; Chueasuwanchai, S. The microscopic agglutination test (MAT) is an unreliable predictor of infecting Leptospira serovar in Thailand. *Am. J. Trop. Med. Hyg.* **2009**, *81*, 695–697. [CrossRef]

97. Mensah, E.; El Zowalaty, M.E. Arboviruses in South Africa, known and unknown. *Future Virol.* **2018**, *13*, 787–802. [CrossRef]

98. Cantas, L.; Suer, K. The important bacterial zoonoses in "one health" concept. *Front. Public Health* **2014**, *2*, 144. [CrossRef]

99. Nicholson, W.L.; Allen, K.E.; McQuiston, J.H.; Breitschwerdt, E.B.; Little, S.E. The increasing recognition of rickettsial pathogens in dogs and people. *Trends Parasitol.* **2010**, *26*, 205–212. [CrossRef] [PubMed]

100. Centers for Disease Control and Prevention. Update: Investigation of bioterrorism-related anthrax and interim guidelines for exposure management and antimicrobial therapy, October 2001. *Morb. Mortal. Wkly. Rep.* **2001**, *50*, 909.

101. Ghosh, R.; Uppal, B.; Aggarwal, P.; Chakravarti, A.; Jha, A.K. Increasing antimicrobial resistance of *Campylobacter jejuni* isolated from paediatric diarrhea cases in a tertiary care hospital of New Delhi, India. *J. Clin. Diagn. Res. JCDR* **2013**, *7*, 247. [CrossRef] [PubMed]
102. Anthony, F.; Acar, J.; Franklin, A.; Gupta, R.; Nicholls, T.; Tamura, Y.; Thompson, S.; Threlfall, E.; Vose, D.; Van Vuuren, M. Antimicrobial resistance: Responsible and prudent use of antimicrobial agents in veterinary medicine. *Rev. Sci. Et Tech.-Off. Int. Des Epizoot.* **2001**, *20*, 829–837. [CrossRef]
103. Cabello, F.C. Heavy use of prophylactic antibiotics in aquaculture: A growing problem for human and animal health and for the environment. *Environ. Microbiol.* **2006**, *8*, 1137–1144. [CrossRef] [PubMed]
104. Casewell, M.; Friis, C.; Marco, E.; McMullin, P.; Phillips, I. The European ban on growth-promoting antibiotics and emerging consequences for human and animal health. *J. Antimicrob. Chemother.* **2003**, *52*, 159–161. [CrossRef] [PubMed]

 *pathogens*

Article

Emergence and Spread of Extended Spectrum β-Lactamase Producing Enterobacteriaceae (ESBL-PE) in Pigs and Exposed Workers: A Multicentre Comparative Study between Cameroon and South Africa

Luria Leslie Founou [1,2,*], Raspail Carrel Founou [1,3], Noyise Ntshobeni [4], Usha Govinden [1], Linda Antoinette Bester [5], Hafizah Yousuf Chenia [4], Cyrille Finyom Djoko [6,7] and Sabiha Yusuf Essack [1]

[1] Antimicrobial Research Unit, School of Health Sciences, College of Health Sciences, University of KwaZulu-Natal, Durban 4000, South Africa; czangue@yahoo.fr (R.C.F.); Govindenu@ukzn.ac.za (U.G.); Essacks@ukzn.ac.za (S.Y.E.)

[2] Department of Food Safety and Environmental Microbiology, Centre of Expertise and Biological Diagnostic of Cameroon (CEDBCAM), Yaoundé 8242, Cameroon

[3] Department of Clinical Microbiology, Centre of Expertise and Biological Diagnostic of Cameroon (CEDBCAM), Yaoundé 8242, Cameroon

[4] Discipline of Microbiology School of Life Sciences, College of Agriculture, Engineering and Sciences, University of KwaZulu-Natal, Durban 4000, South Africa; nb.ntshobeni@gmail.com (N.N.); Cheniah@ukzn.ac.za (H.Y.C.)

[5] Biomedical Resource Unit, School of Laboratory Medicine and Medical Sciences, College of Health Sciences, University of KwaZulu-Natal, Durban 4000, South Africa; Besterl@ukzn.ac.za

[6] Centre for Research and Doctoral Training in Life Science, Health and Environment, The Biotechnology Centre, University of Yaoundé I, Yaoundé 337, Cameroon; cfdjoko@gmail.com

[7] Metabiota Inc., Yaoundé 8242, Cameroon

[*] Correspondence: luriafounou@gmail.com; Tel.: +237-676-54-24-45

Received: 18 December 2018; Accepted: 11 January 2019; Published: 16 January 2019

Abstract: Extended spectrum β-lactamase-producing *Enterobacteriaceae* (ESBL-PE) represent a significant public health concern globally and are recognized by the World Health Organization as pathogens of critical priority. However, the prevalence of ESBL-PE in food animals and humans across the farm-to-plate continuum is yet to be elucidated in Sub-Saharan countries including Cameroon and South Africa. This work sought to determine the risk factors, carriage, antimicrobial resistance profiles and genetic relatedness of extended spectrum β-lactamase producing *Enterobacteriaceae* (ESBL-PE) amid pigs and abattoir workers in Cameroon and South Africa. ESBL-PE from pooled samples of 432 pigs and nasal and hand swabs of 82 humans were confirmed with VITEK 2 system. Genomic fingerprinting was performed by ERIC-PCR. Logistic regression (univariate and multivariate) analyses were carried out to identify risk factors for human ESBL-PE carriage using a questionnaire survey amongst abattoir workers. ESBL-PE prevalence in animal samples from Cameroon were higher than for South Africa and ESBL-PE carriage was observed in Cameroonian workers only. Nasal ESBL-PE colonization was statistically significantly associated with hand ESBL-PE (21.95% vs. 91.67%; $p = 0.000$; OR = 39.11; 95% CI 2.02–755.72; $p = 0.015$). Low level of education, lesser monthly income, previous hospitalization, recent antibiotic use, inadequate handwashing, lack of training and contact with poultry were the risk factors identified. The study highlights the threat posed by ESBL-PE in the food chain and recommends the implementation of effective strategies for antibiotic resistance containment in both countries.

Keywords: antibiotic resistance; Enterobacteriaceae; ESBL; food chain; one health

---

1. Introduction

Enterobacteriaceae are rod-shaped, Gram-negative bacteria, fermenting glucose, usually motile and facultative anaerobes, with the majority of genera being natural residents of gastrointestinal tract of animals, humans and some of them can be found in the environment [1,2]. The extensive use of third and fourth generation cephalosporins in human and animal health, has led to the emergence of extended spectrum β-lactamase-producing *Enterobacteriaceae* (ESBL-PE). ESBL-PE represent a significant public health concern globally and have recently been classified by the World Health Organization as pathogens of critical priority in research [3].

Several studies have detected ESBL-PE in food animals, especially pigs, poultry and cattle and food products throughout the world and their transmission from livestock to humans in the farm-to-plate continuum has been evidenced [4,5]. However, the prevalence of ESBL-PE in food animals and humans across the farm-to-plate continuum is yet to be elucidated in Sub-Saharan countries including Cameroon and South Africa. It is therefore imperative to understand the epidemiology and determine the burden of ESBL-PE in food animals in order to highlight the threat posed by these resistant bacteria and provide evidence for decision-makers to implement effective prevention and containment measures of antibiotic resistance (ABR) in Cameroon and South Africa. The objectives of this study were thus to assess and compare the colonization, antibiotic resistance profiles and genetic relatedness of ESBL-PE among pigs and exposed workers and delineate risk factors of ESBL-PE carriage in humans in these countries.

2. Results

2.1. Demographic Characteristics

Altogether, 114 people were contacted in the five selected slaughterhouses and 83 (73%) workers agreed to participate in the study, with the response rate being higher in Cameroon (71%) than in South Africa (59%). Table 1 describes nasal and hand ESBL-PE carriage of workers in relation to individual, medical/clinical history and slaughterhouse-related characteristics.

Table 1. Nasal and hand ESBL-PE carriage of workers in relation to personal, medical/clinical and slaughterhouse-related characteristics. Of 84 workers enrolled, one withdrew prior to the sample collection and six refused the nasal sampling, yielding a total of 83 hand and 77 nasal samples. A few workers could not recall precise information whilst other refused to answer some questions leading to missing information that was not considered in the analysis.

Variables	Nasal Sample			Hand Sample		
	Frequency n (%)	Prevalence ESBL-PE (%)	Overall *p*-Value	Frequency n (%)	Prevalence ESBL-PE (%)	Overall *p*-Value
Personal characteristics						
Country						
Cameroon	53 (69)	67.92	0.000	53 (64)	79	0.000
South Africa	24 (31)	0		30 (36)	0	
Gender						
Female	9 (12)	44.44	0.883	12 (12)	41.67	0.503
Male	68 (88)	47.06		71 (88)	52.11	
Age						
21–30	31 (40)	41.94		32 (39)	46.88	
31–40	26 (34)	50	0.084	28 (34)	42.86	0.063
41–50	13 (17)	38.46		14 (17)	71.43	
51–60	5 (6)	100		6 (7)	83.33	
Above 60	2 (3)	0		3 (3)	0	

Table 1. *Cont.*

Variables	Nasal Sample			Hand Sample		
	Frequency n (%)	Prevalence ESBL-PE (%)	Overall *p*-Value	Frequency n (%)	Prevalence ESBL-PE (%)	Overall *p*-Value
Educational level						
Illiterate	4 (5)	50		5 (6)	40	
Primary school not achieved	6 (8)	50	0.048	7 (8)	42.86	0.032
Primary school	34 (44)	64.71		35 (42)	71.43	
Secondary school	27 (35)	25.93		27 (33)	37.04	
High school/university	6 (8)	33.33		8 (10)	25	
Average monthly income (US $)						
Below 55	8 (10)	62.50		8 (14)	62.5	
55–110	14 (19)	78.57		14 (29)	85.71	
110–165	12 (16)	66.67	0.007	12 (17)	58.33	0.004
165–220	10 (13)	40		10 (17)	70	
220–275	20 (27)	20		24 (10)	25	
Above 275	11 (15)	27.27		13 (12)	30.77	
Relative working at hospital or with animals						
Yes	42 (55)	64.29	0.001	44 (53)	68.18	0.001
No	35 (45)	25.71		39 (47)	30.77	
Clinical factors						
Recent hospitalization (within one year of sampling)						
Yes	21 (27)	39.29	0.032	21 (25)	71.43	0.027
No	56 (73)	66.67		62 (75)	43.55	
Nasal problem						
Yes	11 (14)	36.36	0.456	11 (13)	45.45	0.714
No	66 (86)	48.48		72 (87)	51.39	
Skin problem						
Yes	14 (18)	28.57	0.132	14 (17)	35.71	0.222
No	63 (82)	50.76		69 (83)	53.62	
Recent antibiotic use (month prior the sampling)						
Yes	38 (49)	55.26	0.140	38 (64)	71.05	0.001
No	39 (51)	38.46		45 (36)	33.33	
Slaughterhouse-related factors						
Closeness of abattoir with house						
Yes	32 (42)	40.63	0.363	14 (33)	17	0.184
No	45 (58)	51.11		28 (67)	34	
Abattoir						
SH001	21 (27)	76.19		21 (25)	85.71	
SH002	19 (25)	36.84	0.000	19 (23)	63.16	0.000
SH003	13 (17)	100		13 (16)	92.31	
SH004	4 (5)	0		10 (12)	0	
SH005	20 (26)	0		20 (24)	0	

Table 1. *Cont.*

Variables	Nasal Sample			Hand Sample		
	Frequency n (%)	Prevalence ESBL-PE (%)	Overall *p*-Value	Frequency n (%)	Prevalence ESBL-PE (%)	Overall *p*-Value
Principal activity or working area						
Slaughterer	34 (44)	58.82		34 (41)	58.82	
Transport of pig/pork	5 (7)	80		5 (6)	80	
Wholesaler	7 (9)	28.57		7 (8)	85.71	
Butcher	5 (7)	80	0.012	5 (6)	80	0.000
Retailer of viscera *	7 (9)	71.43		7 (8)	85.71	
Retailer of grilled pork #	1 (1)	0		1 (1)	100	
Scalding of pigs	3 (4)	0		3 (4)	0	
Evisceration	8 (10)	0		14 (17)	0	
Transport of viscera/blood	1 (1)	0		1 (1)	0	
Veterinarian	5 (7)	20		5 (6)	20	
Meat inspector	1 (1)	0		1 (1)	0	
Training to practice profession						
Yes	28 (36)	3.57	0.000	34 (41)	2.94	0.000
No	49 (64)	71.43		49 (59)	83.67	
Year in profession						
[0–4]	31 (43)	35.48		31 (39)	38.71	
[5–9]	6 (8)	66.67	0.356	8 (10)	50	0.357
[10–14]	22 (30)	50		24 (30)	58.33	
Above 15	14 (19)	57.14		16 (20)	62.50	
Intensity of pig's contact (rare, low, frequent, very frequent)						
Always	35 (45)	51.43		35 (42)	57.14	
Almost always	32 (42)	37.50	0.348	38 (46)	39.47	0.136
Sometimes	10 (13)	60		10 (12)	70	
Contact with other animals during handling or various procedures of processing of animals at the abattoir						
Yes	38 (50)	60.53	0.046	39 (48)	69.23	0.004
No	38 (50)	34.21		42 (52)	35.71	
Intensity of contact with other animals						
Always	8 (21)	87.50		8 (20)	100	
Almost always	9 (24)	22.22	0.025	10 (26)	30	0.006
Sometimes	17 (45)	58.82		17 (44)	70.59	
Rarely	4 (10)	100		4 (10)	100	

* *retailer of viscera*: street-vendor buying pig's viscera from abattoir workers, performing manual cleaning in order to sells ready-to-eat meal; # *retailer of grilled pork*: street-vendor acquiring pork at the slaughterhouse in order to sells ready-to-eat grilled pork.

2.2. ESBL-PE Status in Humans

Out of the 53 workers sampled in Cameroon, 42 (79%) and 36 (68%) were colonized by hand and nasal ESBL-PE, respectively. The main species identified were *E. coli*, *Enterobacter spp.* and *K. pneumoniae* (Supplementary Table S1). In contrast, in South Africa, *Enterobacteriaceae* was not isolated from slaughterhouse workers.

Cameroonian isolates exhibited elevated resistance to ampicillin, trimethoprim-sulfamethoxazole, cefuroxime, cefuroxime-axetil, cefotaxime, ceftazidime and amoxicillin-clavulanic acid (Table 2) with no resistance observed against imipenem, ertapenem, meropenem and tigecycline (Table 2). The profiles AMP.TMP/SXT.CXM.CXM-A.CTX (34%) and AMP.AMC.TZP.CXM.CXM-A.CTX.CAZ.TMP/SXT (7%) were predominant in hand and nasal ESBL-*E. coli*, respectively, in humans in Cameroon (Table 3).

Table 2. Antimicrobial susceptibility results of extended-spectrum β-lactamase-producing *Enterobacteriaceae* (ESBL-PE) isolated from pigs and humans.

Antibiotics	Cameroon				South Africa	
	Pig		Human		Pig	
	MIC (µg/mL) Range	No. (%) Resistant Isolates	MIC (µg/mL) Range	No. (%) Resistant Isolates	MIC (µg/mL) Range	No. (%) Resistant Isolates
Ampicillin	≥32	126 (95)	≤2–≥32	32(73)	≥32	38 (100)
Amoxicillin-clavulanate	4–≥32	54(40)	≤2–≥32	8(18)	8–16	2(5)
Piperacillin-tazobactam	≤4–≥128	24(18)	≤4–64	2(5)	≤4	0
Cefuroxime	4–≥64	124(93)	≤1–≥64	19(43)	≥64	38 (100)
Cefuroxime-axetil	4–≥64	125(93)	≤1–≥64	19(43)	≥64	38 (100)
Cefoxitin	≤4–≥64	10(7)	≤4–≥64	3(7)	≤4	0
Cefotaxime	≤1–≥64	118(88)	≤1–≥64	14(32)	4–≥64	38 (100)
Ceftazidime	≤1–≥64	93(69)	≤1–≥64	8(18)	≤1–4	1 (3)
Cefepime	≤0.5–≥64	6(4)	≤1–≥64	2(5)	≤1–4	1 (3)
Meropenem	≤0.25	0	≤0.25	0	≤0.25	0
Imipenem	≤0.25	0	≤0.25	0	≤0.25	0
Ertapenem	≤0.5	0	≤0.5	0	≤0.5	0
Amikacin	≤2–16	11(8)	≤2–16	1(2)	≤2–16	1 (3)
Gentamicin	≤1–≥16	43(32)	≤1–≥16	3(7)	≤1–≥16	7(18)
Ciprofloxacin	≤0.25–≥4	33(25)	≤0.25–≥4	2(5)	≤0.25	0
Tigecycline	≤0.5–2	0	≤0.5–1	0	≤0.5–1	0
Nitrofurantoin	≤16–64	0	≤16–128	1(2)	≤16–64	0
Colistin	≤0.5	0	≤0.5–4	1(2)	≤0.5–8	1(3)
Trimethoprim-sulfamethoxazole	≤20–≥320	119 (89)	≤20–≥320	22(50)	≤20–≥320	36(95)

Table 3. Antimicrobial resistance profiles of extended-spectrum β-lactamase-producing *Enterobacteriaceae* (ESBL-PE) strains isolated from humans.

Bacteria	Resistance Profiles	Cameroon	
		Nasal, n (%)	Hand, n (%)
E. coli	AMP.AMC.TZP.CXM.CXM-A.CTX.CAZ.TMP/SXT	1 (50)	0
	AMP.CXM.CXM-A.CAZ.CS	0	1(8)
	AMP.TMP/SXT.CXM.CXM-A.CTX	0	4(31)
	AMP.TMP/SXT.CXM.CXM-A.CTX.CAZ.AMC.GM.CIP.FEP	0	1(8)
	AMP.TMP/SXT.CXM.CXM-A.CTX.CAZ.AMC.TZP	0	1(8)
	AMP.TMP/SXT.CXM.CXM-A.CTX	0	1(8)
	AMP.CXM.CXM-A.CTX.FEP.TMP/SXT	0	1(8)
E. dissolvens	AMP.AMC.CXM.CXM-A.FOX.CTX.TMP/SXT	1 (50)	0
S. sonnei	AMP.TMP/SXT.CXM.CXM-A.CTX.CAZ	0	1(8)
K. pneumoniae	AMP.AMC.CXM.CXM-A.CTX.TMP/SXT.GM.FT	0	1(8)
	AMP.AMC.CXM.CXM-A.CTX.CAZ.AN.GM.CIP	0	1(8)
	AMP.TMP/SXT.CXM.CXM-A.CTX.CAZ	0	1(8)
	Grand Total	2 (100)	13 (100)

AMP: Ampicillin; AMC: Amoxicillin-clavulanate; TZP: Piperacillin-tazobactam; CXM: Cefuroxime; CXM-A: Cefuroxime-Acetyl; CTX: Cefotaxime; CAZ: Ceftazidime; TMP/SXT: Trimethoprim-Sulfamethoxazole; FOX: Cefoxitin; GN: Gentamicin; CIP: Ciprofloxacin; FEP: Cefepime; CS: Colistin.

2.3. Epidemiological Background of ESBL-PE in Pigs

Overall, the prevalence of ESBL-PE in the pooled nasal and rectal samples was 75% (108/144) and 71% (102/144), respectively (Table 4). At country-level, 42% (30/72) and 50% (36/72) ESBL-PE were detected in rectal and nasal pooled samples in South Africa respectively, whereas a 100% ESBL-PE prevalence was isolated in both specimen types in Cameroon (Table 1). In Cameroon, the main species identified were *E. coli* (61%) and *Klebsiella pneumoniae* (25%) whereas in South Africa, *E. coli* was the sole *Enterobacteriaceae* species isolated in both types of samples (Supplementary Table S2).

Table 4. Extended-spectrum β-lactamase-producing *Enterobacteriaceae* (ESBL-PE) in pooled nasal and rectal samples.

Characteristics	Nasal Samples			Rectal Samples		
	Frequency Pooled Samples, n (%)	Nasal ESBL, n (%)	Overall *p*-Value	Frequency Pooled Samples, n (%)	Rectal ESBL, n (%)	Overall *p*-Value
Country						
Cameroon	72 (50)	72 (100)	0.000	72 (50)	72 (100)	0.000
South Africa	72 (50)	36 (50)		72 (50)	30 (41.67)	
Abattoir						
SH001	43 (30)	43 (100)		43 (30)	43 (100)	
SH002	19 (13)	19 (100)	0.000	19 (13)	19 (100)	0.000
SH003	10 (7)	10 (100)		10 (7)	10 (100)	
SH004	40 (28)	19 (47.50)		40 (28)	9 (22.50)	
SH005	32 (22)	17 (53.13)		32 (22)	21 (65.63)	
Gender						
Sow	79 (55)	64 (81.01)	0.066	79 (55)	59 (74.68)	0.262
Boar	65 (45)	44 (67.69)		65 (45)	43 (66.15)	
Time point						
First	42 (29)	31 (73.81)		42 (29)	34 (80.95)	
Second	54 (38)	45 (83.33)	0.149	54 (38)	40 (74.07)	0.050
Third	48 (33)	32 (66.67)		48 (33)	28 (58.33)	

ESBL-PE isolated from pigs exhibited high resistance to ampicillin, cefuroxime, cefuroxime-acetyl, cefotaxime, ceftazidime and trimethoprim-sulfamethoxazole in both countries (Table 2). One South African isolate expressed high resistance to colistin (8 mg/L) and no resistance to ertapenem, meropenem, imipenem and tigecycline was observed. The majority of ESBL-producing *E. coli* isolated from pigs in both countries showed the resistance profile AMP.TMP/SXT.CXM.CXM-A.CTX in both type of samples (Table 5).

2.4. Genotypic Relatedness

ERIC-PCR allowed the differentiation of the 93 *E. coli* into 14 clusters named alphabetically from A-N (Figure 1). A batch of isolates in cluster M (PR210, PR212E *, PR209E2, PR246B1C and PN254E), collected from pigs of abattoir SH004 and SH005 in South Africa was considered to be closely related. Moreover, great interest was observed in cluster I, where one pair of animal strains, PR085E3 and PR209E1 isolated in abattoirs SH002 and SH004 in Cameroon and South Africa, respectively, showed 100% similarity and were closely related with a human strain (HN503E2II) detected in abattoir SH001 in Cameroon (Figure 1).

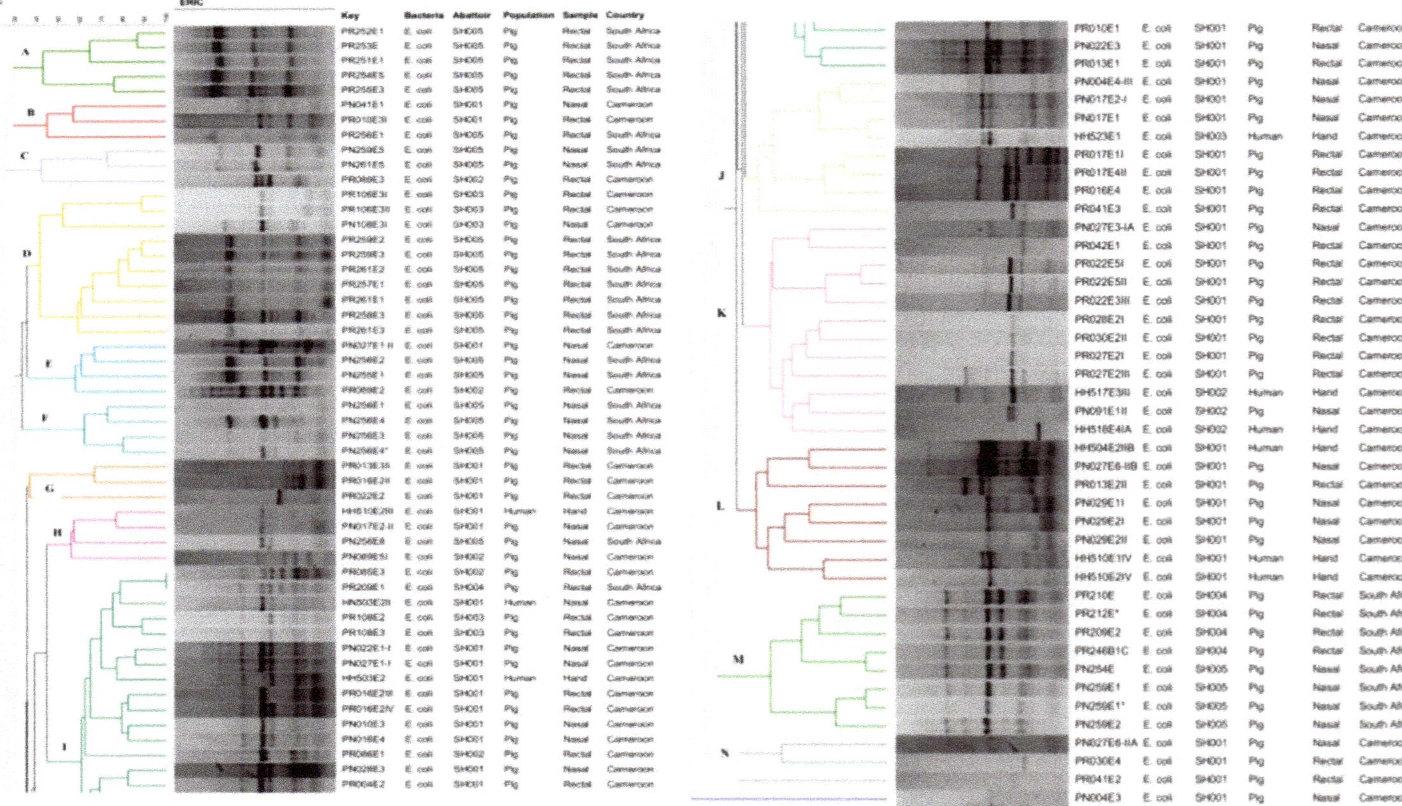

Figure 1. Genotypic relationship of ESBL-*E. coli* strains (n = 93) detected from exposed workers and pigs in Cameroon and South Africa. Dendogram generated by Bionumerics using UPGMA method and the Dice similarity coefficient.

2.5. Risk Factors of Human ESBL-PE Carriage

Table 6 shows the relationship between ESBL-PE carriage in workers and the foremost putative risk factors. Nasal and hand ESBL-PE colonization were univariately associated with an odds ratio (OR) of 39.11 (95% CI 2.02–755.72; $p = 0.015$). Other determinants, univariately associated with nasal and hand ESBL-PE carriage were previous hospitalization, recent antibiotic use, inadequate handwashing, occupation of relatives and year in the employment. The multivariate analysis reveals that nasal and hand ESBL-PE carriage in humans were associated with contact with other animals, particularly poultry with high statistical significance for both sample types (OR = 5.83, 95% CI 1.58–21.48, $p = 0.008$; vs. OR = 8.41, 95% CI 2.27–31.11, $p = 0.001$).

Table 5. Antimicrobial resistance profiles of extended-spectrum β-lactamase-producing *Enterobacteriaceae* (ESBL-PE) detected from pigs.

Bacteria	Resistance Profiles	No. Antibiotics	No. Classes	Cameroon		South Africa	
				Nasal, n (%)	Rectal, n (%)	Nasal, n (%)	Rectal, n (%)
E. coli	AMP.TMP/SXT.CXM.CXM-A.CTX	5	2	3 (5)	2 (4)	0	29(94)
	AMP.TMP/SXT.CXM.CXM-A.CTX.CAZ	6	2	7 (12)	11 (20)	0	0
	AMP.TMP/SXT.CXM.CXM-A.CTX.CAZ.FEP	7	2	2 (3)	2 (4)	0	0
	AMP.TMP/SXT.CXM.CXM-A.CTX.CAZ.GM.CIP	8	4	1 (2)	2 (4)	0	0
	AMP.TMP/SXT.CXM.CXM-A.CTX.CAZ.CIP	7	3	1 (2)	0	0	0
	AMP.TMP/SXT.CXM.CXM-A.CTX.CAZ.AMC	8	3	1 (2)	3 (5)	0	0
	AMP.TMP/SXT.CXM.CXM-A.CTX.CAZ.AMC.GM.CIP	9	4	2 (3)	3 (5)	0	0
	AMP.TMP/SXT.CXM.CXM-A.CTX.CAZ.AMC.TZP	8	2	5 (8)	5 (9)	0	0
	AMP.TMP/SXT.CXM.CXM-A.CTX.CAZ.AMC.TZP.FOX.FEP.GM.CIP	12	4	1 (2)	0	0	0
	AMP.CXM.CXM-A.CTX.CAZ	5	1	0	2 (4)	0	0
	AMP.TMP/SXT.CXM.CXM-A.CTX.CAZ.FEP.GM.	8	3	0	1 (2)	0	0
	AMP.TMP/SXT.CXM.CXM-A.CTX.CIP	6	3	0	1 (2)	0	0
	AMP.TMP/SXT.CXM.CXM-A.CTX.CAZ.CIP.AMC	8	3	0	5 (9)	0	0
	AMP.TMP/SXT.CXM.CXM-A.CTX.GM	6	3	0	1 (2)	0	0
	AMP.TMP/SXT.CXM.CXM-A.CTX.CAZ.CIP.AMC.TZP	9	3	0	1 (2)	0	0
	AMP.TMP/SXT.CXM.CXM-A.CTX.CAZ.CIP.GM.TZP.FEP	10	4	0	1 (2)	0	0
	AMP.CXM.CXM-A.CTX	4	1	0	0	0	2 (6)
	AMP.CXM.CXM-A.CTX.TMP/SXT.CAZ.FEP.AK.GM.CS	10	4	0	0	1 (14)	0
	AMP.CXM.CXM-A.CTX.TMP/SXT.GM	6	3	0	0	1 (14)	0
	AMP.CXM.CXM-A.CTX.TMP/SXT.GM.AMC	7	3	0	0	5 (71)	0
	AMP.TMP/SXT.CXM.CXM-A.CTX.CAZ.CIP.GM.AMC.TZP	10	4	0	2 (4)	0	0
K. pneumoniae	AMP.TMP/SXT.CXM.CXM-A.CTX.GM	6	3	2 (3)	1 (2)	0	0
	AMP.TMP/SXT.CXM.CXM-A.CTX.CAZ.AK.GM.CIP.AMC	10	4	0	6 (11)	0	0
	AMP.TMP/SXT.CXM.CXM-A.CAZ.AMC.TZP	7	2	0	1 (2)	0	0
	AMP.TMP/SXT.CXM.CXM-A.CTX.CAZ.GM.AMC.TZP	9	3	4 (7)	2 (4)	0	0
	AMP.TMP/SXT.CXM.CXM-A.CTX.CAZ.AMC.TZP.AK.GM.CIP	11	4	0	3(5)	0	0
	AMP.TMP/SXT.CXM.CXM-A.CTX	5	2	1 (2)	0	0	0
	AMP.TMP/SXT.CXM.CXM-A.CTX.CAZ	6	2	4 (7)	0	0	0
	AMP.TMP/SXT.CXM.CXM-A.CTX.CAZ.GM	7	3	8 (14)	0	0	0
	AMP.TMP/SXT.CXM.CXM-A.CTX.CAZ.GM.AMC.TZP	9	3	4 (7)	0	0	0
K. ozaenae	AMP.TMP/SXT.CXM.CXM-A.CTX.CAZ.AK.GM.CIP.AMC	10	4	1 (2)	0	0	0
Enterobacter cloacae	AMP.AMC.CXM.CXM-A.FOX.CTX.TMP/SXT	7	2	4 (7)	0	0	0
C. freundii	AMP.AMC.CXM.CXM-A.FOX.CTX.TMP/SXT	7	2	3 (5)	0	0	0

Table 5. *Cont.*

Bacteria	Resistance Profiles	No. Antibiotics	No. Classes	Cameroon		South Africa	
				Nasal, n (%)	Rectal, n (%)	Nasal, n (%)	Rectal, n (%)
S. sonnei	AMP.CTX.CAZ.TMP/SXT	4	2	1 (2)	0	0	0
	AMP.TMP/SXT.CXM.CXM-A.CTX.CAZ	6	2	4 (7)	0	0	0

AMP: Ampicillin; AMC: Amoxicillin-clavulanate; AK: Amikacin; CXM: Cefuroxime; CXM-A: Cefuroxime-acetyl; CTX: Cefotaxime; CAZ: Ceftazidime; CS: Colistin; CIP: Ciprofloxacin; FEP: Cefepime; FOX: Cefoxitin; GM: Gentamicin; TMP/SXT: Trimethoprim-sulfamethoxazole; TZP: Piperacillin-tazobactam.

Table 6. Predictors of nasal and hand ESBL-PE carriage among humans. Univariate and multivariate analysis (logistic regression).

Variables	Univariate Logistic Regression Analysis				Multivariate Logistic Regression Analysis			
	Nasal ESBL-PE Carriage		Hand ESBL-PE Carriage		Nasal ESBL-PE Carriage		Hand ESBL-PE Carriage	
	OR (95% CI)	*p*	OR (95% CI)	*p*	OR (95% CI)	*p*	OR (95% CI)	*p*
Abattoir	0.43 (0.22–0.85)	**0.014**	0.28 (0.15–0.55)	**0.000**	2.54 (0.47–17.75)	0.254	1.75 (0.19–15.54)	0.615
Gender	1.11 (0.10–12.39)	0.932	1.52 (0.20–11.38)	0.682	8.74 (1.03–74.16)	**0.047**	27.94 (1.68–463.05)	**0.020**
Educational level	0.61 (0.39–0.94)	**0.024**	0.72 (0.56–0.91)	**0.006**				
Monthly Income	0.57 (0.36–0.89)	**0.014**	0.60 (0.36–0.99)	**0.045**	0.58 (0.35–0.97)	**0.039**	0.76 (0.44–1.31)	0.324
Training	0.01 (0.0003–0.79)	**0.038**	0.006 (0.0005–0.0658)	**0.000**	0.004 (0.00009–0.22)	**0.006**	0.0008 (0.000008–0.09)	**0.003**
Principal Activities	0.63 (0.50–0.78)	**0.000**	0.63 (0.46–0.87)	**0.004**				
Occupation of relative [a]	5.2 (1.46–18.56)	**0.011**	4.82 (0.64–36.56)	0.128	5.62 (1.02–30.82)	**0.047**	3.58 (0.57–22.43)	0.172
Year in Profession	1.40 (0.66–2.97)	0.387	1.35 (0.68–2.69)	0.398				
Age	1.09 (0.80–1.48)	0.595	1.07 (0.61–1.89)	0.817				
Recent hospitalization [b]	3.09 (1.26–7.59)	**0.014**	3.24 (1.18–8.86)	**0.022**	1.28 (0.24–6.87)	0.769	0.57 (0.08–4.12)	0.576
Recent antibiotic use [c]	1.97 (0.40–9.73)	0.402	4.91 (1.20–20.03)	**0.027**				
Skin problem	0.39 (0.17–0.89)	**0.025**	0.48 (0.21–1.08)	0.076				
Nasal problem	0.61 (0.29–1.28)	0.192	0.79 (0.34–1.82)	0.578				
Protective working clothes	0.04 (0.002–0.812)	**0.036**	0.022 (0.002–0.258)	**0.002**				
Inadequate Handwashing	4.71 (2.28–9.70)	**0.000**	3.9 (1.01–15.01)	**0.048**				
Convenient handwashing	0.08 (0.017–0.41)	**0.002**	0.04 (0.013–0.145)	**0.000**				
Intensity of contact with pigs	0.97 (0.50–1.87)	0.920	0.96 (0.40–2.31)	0.934				
Contact with other animals	2.95 (0.87–10.04)	0.084	4.05 (1.42–11.53)	**0.009**				
Contact with poultry	5.83 (1.58–21.48)	0.008	8.41 (2.27–31.11)	**0.001**	9.93 (1.37–71.63)	**0.023**	24.22 (1.28–457.35)	**0.034**
Pig colonization Nasal ESBL (yes or No)	1.04 (0.93–1.16)	0.509	1.06 (0.95–1.17)	0.313				
Pig colonization Rectal ESBL (yes or No)	1.03 (0.93–1.15)	0.585	1.05 (0.96–1.16)	0.273				

[a]: Relative working with food animals, food products or at hospital, [b]: Within one year prior the sampling date; c: Within one month prior the sampling date.

3. Discussion

Enterobacteriaceae and especially ESBL-PE, were acknowledged as critical priority antibiotic-resistant bacteria (ARB) by the WHO and their emergence at the animal-human-environment interface presents a to serious and multifaceted public health concern globally [3]. This study investigated the carriage, risk factors, antibiotic resistance profiles and genetic relatedness of ESBL-PE isolated from apparently healthy pigs and occupationally exposed workers in Cameroon and South Africa.

The overall prevalence of human ESBL-PE carriage was 50% in hand and 45.75% in nasal samples. Comparable data was reported by Magoue et al. (2013) in Cameroon, where the prevalence of ESBL-PE faecal carriage was 45% in outpatients in the region of Adamaoua [6]. Our findings are nevertheless higher than that described by Dohmen et al. (2015) where a 27% prevalence of ESBL-PE carriage in faecal samples of people with daily exposure to pigs in Netherlands was described [7].

Our results are in contrast to a study of Fisher et al., (2016), where none of the 66.7% *Enterobacteriaceae* detected in the nares of participants were ESBL producers and where the authors concluded that nares were a negligible reservoir for colonization of ESBL-PE in pig's exposed workers [8]. Our finding shows that the prevalence of ESBL-PE carriage in nasal samples substantially increased (8.33 vs. 91.67%; $p < 0.001$) and was statistically significantly correlated with their carriage on hand (OR 39.11; 95% CI 2.02–755.72; $p = 0.015$). In addition, nasal ESBL-PE carriage was associated with inappropriate handwashing with high statistical significance (OR 4.71; 95% CI 2.28–9.70; $p < 0.001$). This suggests, that nares might likely become reservoir of ESBL-PE when limited hygienic conditions prevail and biosecurity measures are not adequately implemented. It further reveals that, as with the transmission of nosocomial infections in hospital settings, hands constitute important vectors of ABR transmission in the food production industry and may not only drive the transfer from person-to-person but also the contamination of food products intended for the end consumer. Nasal ESBL-PE carriage reported herein might also be ascribed to airborne contamination as recently reported by Dohmen et al., (2017) who revealed that human CTX-M-gr1 carriage was statistically associated with presence of CTX-M-gr1 in dust (OR = 3.5, 95% CI = 0.6–20.9) and that inhalation of air might constitute another transmission route of ESBL-PE in the food chain [9].

The difference in the prevalence of ESBL-PE carriage in humans in both countries could be explained by the fact that South Africa has existing abattoir regulations in place and South African abattoirs were compliant with international food safety standard ISO 22000 and Hazard Analysis Critical Control Points (HACCP) plans. In Cameroon, slaughterhouse/markets were principally low-grade, lacking in basic amenities, with sub-optimal sanitary conditions and limited or non-existent biosecurity measures. The Food and Agriculture Organization for the United Nations (FAO) report on abattoir facilities in Central African countries including Cameroon, already underlined the gaps in term of biosecurity measures in these settings [10]. Our findings, therefore, reinforce the importance of and the need to implement strict biosecurity procedures as when effective prevention and containment measures are implemented, the risk of ABR dissemination is reduced.

The overall prevalence of ESBL-PE in pigs was 71% and 75% in rectal and nasal pooled samples, respectively. The results are consistent with that reported by Le et al., (2015) in food animals and products in Vietnam where a 68.4% prevalence of ESBL-producing *E. coli* was described [11]. They are however lower than that reported in pig farms in Germany, where 88.2% of ESBL-producing *E. coli* was detected [12] and higher than that reported in two other studies with prevalence ranging from 8.6 to 63.4% in food animals and food products in Netherlands, [13] and 8.4% in cattle in Switzerland [2].

The high rate of ESBL-PE carriage detected in both nasal and rectal samples in Cameroon may suggest that ESBL-PE are consistently widespread in food animals in Cameroon, disseminate in the farm-to-plate continuum and represent a grave public health threat in the country. Similarly, the ESBL-PE prevalence detected in pigs in South Africa is not surprising, especially because the use of antibiotics as growth promoter agents is legally allowed in the country [14]. These findings reveal gaps in the current state of knowledge about antibiotic use and ABR in food animals and suggests that the debate about ABR-related consequences in the farm-to-plate continuum is neglected in Cameroon and South Africa and should be more seriously considered in these countries. Additionally, our study revealed a high frequency (95%) of ESBL-producing *E. coli*, emphasizing the relevance of this indicator bacteria as a serious public health issue.

ERIC analysis demonstrated relative associations amongst human and animal isolates within and across countries. Some strains isolated in humans were highly related to those detected from pigs at similar or dissimilar abattoirs suggesting that the occurrence of ESBL-PE in humans may have an animal origin or vice-versa and that these bacteria may spread to humans via the food chain, allowing their dissemination to the global population. Although not providing evidence on the transmission dynamics of ESBL-PE, our results nevertheless show an epidemiological link amongst isolates from humans and animals.

Hospitalization, antibiotic use and contact with (food) animals are known risk factors for human ESBL-PE carriage [15]. Twenty-one abattoir workers or their family members had been admitted to a hospital within the year of the sampling. Of these, 39.29% evidenced nasal ESBL-PE carriage and 71.43% hand ESBL-PE colonization (Table 1). Likewise, the majority of workers who had used antibiotics within the month of the sampling were colonized by ESBL-PE in nares (55.26%) and hands (71.05%) (Table 1).

There are certain limitations to consider in this cross-sectional study. First, the duration of ESBL-PE carriage was not investigated and there was no apparent relationship between human ESBL-PE carriage and contact with ESBL-PE colonized pigs (Table 6). Secondly, in contrast, a clear association was established between contact with other (food) animals, mainly poultry and human ESBL-PE colonization, with high statistical significance (Table 6), suggesting that further work should be undertaken in high risk populations and other food animals such as poultry in order to expand our understanding on the public health impact of the likely zoonotic transmission of ESBL-PE through the farm-to-plate continuum. Thirdly, the small human sample size precluded any direct conclusions on the prevalence of antibiotic resistance among abattoir workers. Finally, the molecular analyses were only carried out on a representative sub-sample and not all isolates due to financial constraints. Comprehensive molecular analysis would have certainly allowed better understanding of the genetic exchanges and evolution that are likely to occur within and between bacteria in this continuum.

To the best of our knowledge, this is the first report of ESBL-PE in animals and humans in both Cameroon and South Africa taking food safety perspective. The high prevalence of ESBL-PE found in pigs in both countries as well as in humans in Cameroon highlights the food safety issue associated with their presence in the farm-to-plate continuum. It demonstrates the urgent need to implement multi-sectorial, multi-faceted and sustainable collaboration and activities among all stakeholders involved in this continuum in order to reduce the prevalence and contain the dissemination of ESBL-PE and ABR in these countries.

4. Methods

4.1. Study Design and Study Sites

From March to October 2016, a multicentre study was conducted in five abattoirs in Cameroon (n = 3) and South Africa (n = 2). All abattoirs were coded for ethical reasons as SH001, SH002, SH003, SH004 and SH005. They were visited thrice at different time points to allow a representative sample.

4.2. Ethical Considerations

Prior to the implementation of the study, ethical approvals were obtained from the National Ethics Committee for Research in Human Health of Cameroon (Ref. 2016/01/684/CE/CNERSH/SP), Biomedical Research Ethics Committee (Ref. BE365/15) and Animal Research Ethics Committee (Ref. AREC/091/015D) of the University of KwaZulu-Natal. In addition, ministerial approvals from the Cameroonian Ministry of Scientific Research and Innovation (Ref. 015/MINRESI/B00/C00/C10/C14) and

Ministry of Livestock, Fisheries and Animal Industries (Ref. 061/L/MINEPIA/SG/DREPIA/CE) were also granted. This study was further placed on record with the South African National Department of Agriculture, Forestry and Fisheries [Reference: 12/11/1/5 (878)].

4.3. Sampling Procedures and Survey

4.3.1. Animal Sampling Procedure

Apparently healthy and freshly slaughtered/stunned pigs were randomly sampled in both Cameroon and South Africa. The interior cavity of both anterior nares were swabbed and rectal swabs of pigs were obtained using sterile Amies swabs (Copan Italia Spa, Brescia, Italia). Altogether, 432 nasal and rectal pigs were collected in both countries, with the number of samples from each slaughterhouse (SH001, n = 129; SH002, n = 57; SH003, n = 30; SH004, n = 120; SH005, n = 96) proportional to the annual pig production per site.

4.3.2. Human Sampling Procedure

Total sampling was employed where all exposed workers ($\geq$21 years old) willing to participate were recruited in the study upon oral and written informed consent. Participants were requested to answer a questionnaire describing socio-demographic and medical/clinical history, as well as probable risk factors associated with ESBL-PE emergence/colonization and spread. Amies swab was used to collect both anterior nares and hand (between fingers for each right and left hand) samples which were processed within 4 h of collection.

4.4. Bacteriological Analysis

For the bacteriological analysis, three individual pig samples were pooled per abattoir, gender, specimen and area of breeding leading to 288 pools (144 nasal and 144 rectal) representing 432 original specimens collected from 432 pigs. Pooled pig samples and human swabs were cultured onto an in-house selective MacConkey agar supplemented with 2 mg/L cefotaxime (MCA+CTX) and incubated for 18–24 h at 37 °C for ESBL-PE screening. Presumptive ESBL-PE were phenotypically confirmed with Vitek® 2 System (BioMérieux, Marcy l'Etoile, France).

4.5. ESBL Detection, Species Identification and Antimicrobial Susceptibility Testing

Each colony with a unique morphotype growing on MCA+CTX was screened for ESBL production through the standard double disk synergy test (DDST) as recommended by the Clinical Laboratory and Standards Institute (CLSI) [16].

A panel of 19 antibiotics including amoxicillin + clavulanic acid, ampicillin, cefuroxime, cefuroxime axetil, cefoxitin, cefotaxime, ceftazidime, cefepime, imipenem, ertapenem, meropenem, amikacin, gentamicin, ciprofloxacin, tigecycline, piperacillin/tazobactam, nitrofurantoin, colistin and trimethoprim-sulfamethoxazole, were tested using Vitek® 2 System and Vitek® 2 Gram Negative Susceptibility card (AST-N255) (BioMérieux, Marcy l'Etoile, France). The CLSI was used for interpretation of the results excepted for colistin, piperacillin/tazobactam, amoxicillin + clavulanic acid and amikacin that were interpreted using EUCAST breakpoints [17]. *E. coli* ATCC 25922 was used as the control.

4.6. Genotypic Relatedness Determination of ESBL-Producing Escherichia coli

The Thermo Scientific® GeneJet Genomic DNA purification kit (Thermo Fisher Scientific, Johannesburg, South Africa) was used for the genomic DNA extraction following the manufacturer's instructions. ERIC-PCR was carried out using the primers ERIC 1 5′-ATG TAA GCT CCT GGG GAT TCA C-3′ and ERIC 2 5′-AAG TAA GTG ACT GGG GTG AGC G-3′ [18]. Reactions were performed in a 10 µL final solution containing 5 µL DreamTaq Green Polymerase Master Mix 2× (Thermo Fisher Scientific, South Africa), 2.8 µL nuclease free water, 0.1 µL of each primer (100 µM) and 2 µL DNA template and run in an Applied Biosystems 2720 programmable thermal cycler (Thermo Fisher Scientific, Johannesburg, South Africa). The ERIC-PCR protocol implemented included 3 min of initial denaturation at 94 °C, followed by 30 cycles consisting of a denaturation at 94 °C for 30 s, annealing at 50 °C for 1 min, extension at 65 °C for 8 min, a final extension at 65 °C for 16 min and final storage at 4 °C. ERIC profiles were digitized and analysed using Bionumerics (version 7.6, Applied Maths, Austin, TX, USA). The similarity between each strain was assessed using Dice coefficient and dendrograms were constructed using the Unweighted Pair-Group Method Algorithm (UPGMA).

4.7. Data Analysis

Data was encoded and entered into Epi Info (version 7.2, CDC, Atlanta, GA, USA) and Excel (Microsoft Office 2016) and analysed using STATA (version 14.0, STATACorp LLC, College Statioon, TX, USA). A data set was designed for specific human results and, combined animal and abattoir data. Abattoirs were classified as ESBL-positive if an ESBL-PE was identified from at least one pool (nasal or rectal samples). Likewise, each human was categorized as carrier or non-carrier, with carrier being defined as having ESBL-PE in at least one site (nares or hand).

The ESBL-PE prevalence was compared between categories with the chi square test ($p < 0.05$). The relationship between ESBL-PE carriage in pigs and humans was ascertained using logistic regression analyses adjusted for clustering at abattoir level. Likewise, risk factors for ESBL-PE carriage were determined univariately and selected for multivariate analysis when the p-value was <0.2. The McFadden's pseudo R^2 statistic (maximum likelihood method) was used to check the model fit and the final model included all determinants for which the pseudo R^2 was the most elevated with $p < 0.05$ for each dependent variable.

Supplementary Materials: The following are available online at http://www.mdpi.com/2076-0817/8/1/10/s1. Table S1: Overall prevalence of extended spectrum β-lactamase (ESBL) producing bacteria isolated from humans per country and specimen type. Table S2: Overall prevalence of extended spectrum β-lactamase (ESBL) producing bacteria isolated from animals per country and specimen type. Table S3: Prevalence and distribution of extended-spectrum β-lactamase producing-E. coli clusters per abattoir.

Author Contributions: L.L.F. co-developed the study, performed sample collection, microbiological, genomic and statistical analyses, prepared tables and figures and drafted the manuscript. R.C.F. undertook sample collection, laboratory analyses, contributed to data analysis, vetted the results and reviewed the manuscript. N.N. participated in the genomic extraction and fingerprinting analysis. U.G. and L.A.B contributed materials and reagents for sample collection and primary laboratory analyses and took part in the design of the study. H.Y.C. contributed materials and reagents for and vetting of genomic fingerprinting results. C.F.D. contributed equipment, materials and reagents for the sample collection and primary laboratory analyses, coordinated the field implementation in Cameroon, took part in the design of the study and reviewed the manuscript. S.Y.E. co-developed the study, vetted the results and undertook critical revision of the manuscript. All authors read and approved the manuscript.

Funding: L.L. Founou and R.C. Founou are funded by the Antimicrobial Research Unit (ARU) and College of Health Sciences (CHS) of the University of KwaZulu-Natal. The National Research Foundation funded this study through the DST/NRF South African Research Chair in Antibiotic Resistance and One Health (Grant No. 98342), the NRF Competitive Grant for Rated Researchers (Grant No.: 106063) and the NRF Incentive Funding for Rated Researchers (Grant No. 85595), awarded to S.Y. Essack. Any opinions, results and conclusions or recommendations expressed in this review are those of the authors and therefore do not represent the official position of the funders. The funders had no role in the study design, preparation of the manuscript nor the decision to submit the work for publication.

Acknowledgments: The authors express their gratitude to the Cameroonian Ministry of Livestock, Fisheries and Animal Industries as well as to the Cameroonian Ministry of Scientific Research and Innovation of Cameroon for the study approval and strong support during the field implementation in the country. Metabiota Cameroon Limited, the Military Health Research Centre (CRESAR) and the Laboratory for Public Health Biotechnology/The Biotechnology Center of the University of Yaoundé I through Professor Wilfred Mbacham, are also acknowledged for their administrative, logistical and cold chain support during the sampling and baseline analysis stages. The National Health Laboratory Service (NHLS), especially Mlisana Koleka, Sarojini Govender and Thobile Khanyile are acknowledged for the assistance with the phenotypic identification and MIC determination. Keith Perret, KwaZulu-Natal Veterinary Services is truly appreciated for facilitating the administrative procedure pertaining to the sample collection in South Africa. The Drug Delivery Research Unit of the University of KwaZulu-Natal, through Thirumala Govender and Chunderika Mocktar is thanked for the cold chain support. The authors show their gratitude to Serge Assiene, Arthur Tchapet and Zamabhele Kubone, for their assistance with the sample collection and preliminary screening of samples in Cameroon and South Africa, respectively. We are very thankful to the abattoir owners/coordinators in South Africa, abattoirs' leaders, supervisors, food safety inspectors, veterinarians, workers and study participants, for their support and assistance during the sample collection in both Cameroon and South Africa.

Conflicts of Interest: Professor Sabiha Essack is chairperson of the Global Respiratory Infection Partnership, sponsored by an unrestricted educational grant from Reckitt and Benckiser (Pty.) Ltd. UK.

References

1. Aidara-Kane, A.; Andremont, A.; Collignon, P. Antimicrobial resistance in the food chain and the AGISAR initiative. *J. Infect. Public Health* **2013**, *6*, 162–165. [CrossRef] [PubMed]

2. Reist, M.; Geser, N.; Hächler, H.; Schärrer, S.; Stephan, R. ESBL-producing *Enterobacteriaceae*: Occurrence, risk factors for fecal carriage and strain traits in the swiss slaughter cattle population younger than 2 years sampled at abattoir level. *PLoS ONE* **2013**, *8*, e71725. [CrossRef] [PubMed]

3. World Health Organization (WHO). *Global Priority List of Antibiotic Resistant Bacteria to Guide Research, Discoveries and Development of New Antibiotics*; WHO: Geneva, Switzerland, 2017.

4. Founou, L.L.; Founou, R.C.; Essack, S.Y. Antibiotic resistance in the food chain: A developing country-perspective. *Front. Microbiol.* **2016**, *7*, 1881. [CrossRef] [PubMed]

5. Ewers, C.; Bethe, A.; Semmler, T.; Guenther, S.; Wieler, L.H. Extended-spectrum β-lactamase-producing and AmpC-producing *Escherichia coli* from livestock and companion animals, and their putative impact on public health: A global perspective. *Clin. Microbiol. Infect.* **2012**, *18*, 646–655. [CrossRef] [PubMed]

6. Magoué, C.L.; Melin, P.; Gangoué-Piéboji, J.; Okomo Assoumou, M.C.; Boreux, R.; De Mol, P. Prevalence and spread of extended-spectrum β-lactamase-producing *Enterobacteriaceae* in Ngaoundere, Cameroon. *Clin. Microbiol. Infect.* **2013**, *19*, 416–420. [CrossRef] [PubMed]

7. Dohmen, W.; Bonten, M.J.; Bos, M.E.; van Marm, S.; Scharringa, J.; Wagenaar, J.A.; Heederik, D.J. Carriage of extended-spectrum β-lactamases in pig farmers is associated with occurence in pigs. *Clin. Microbiol. Infect.* **2015**, *21*, 917–923. [CrossRef] [PubMed]

8. Fischer, J.; Hille, K.; Mellmann, A.; Schaumburg, F.; Kreienbrock, L.; Köck, R. Low-level antimicrobial resistance of *Enterobacteriaceae* isolated from the nares of pig-exposed persons. *Epidemiol. Infect.* **2016**, *144*, 686–690. [CrossRef] [PubMed]

9. Dohmen, W.; Schmitt, H.; Bonten, M.; Heederik, D. Air exposure as a possible route for ESBL in pig farmers. *Environ. Res.* **2017**, *155*, 359–364. [CrossRef] [PubMed]

10. Food and Agriculture Organization for the United Nations (FAO). *Étude sur les Abattoirs D'animaux de Boucherie en Afrique Centrale (Cameroun–Congo–Gabon–Tchad)*; FAO: Rome, Italy, 2013.

11. Le, H.V.; Kawahara, R.; Khong, D.T.; Tran, H.T.; Nguyen, T.N.; Pham, K.N.; Jinnai, M.; Kumeda, Y.; Nakayama, T.; Ueda, S.; et al. Widespread dissemination of extended-spectrum β-lactamase-producing, multidrug-resistant *Escherichia coli* in livestock and fishery products in vietnam. *Int. J. Food Contam* **2015**, *2*, 17. [CrossRef]

12. Dahms, C.; Hübner, N.O.; Kossow, A.; Mellmann, A.; Dittmann, K.; Kramer, A. Occurrence of ESBL-producing *Escherichia coli* in livestock and farm workers in mecklenburg-Western Pomerania, Germany. *PLoS ONE* **2015**, *10*, e0143326. [CrossRef] [PubMed]

13. Geser, N.; Stephan, R.; Hachler, H. Occurrence and characteristics of extended-spectrum β-lactamase (ESBL) producing *Enterobacteriaceae* in food producing animals, minced meat and raw milk. *BMC Vet. Res.* **2012**, *8*, 21. [CrossRef] [PubMed]

14. Department of Agriculture, Forestry and Fisheries, South Africa (DAFF). *Fertilizer, Farm Feeds, Agricultural Remedies and Stock Remedies Act, 1947*; Publication of Farm Feeds (Animal Feeds) Policy for Public Comments. Act No. 36 of 1947 C.F.R.; Department of Agriculture, Forestry and Fisheries, South Africa (DAFF): Pretoria, South Africa, 1996. Available online: http://www.daff.gov.za/ (accessed on 21 August 2017).

15. Ben-Ami, R.; Rodríguez-Baño, J.; Arslan, H.; Pitout, J.D.; Quentin, C.; Calbo, E.S.; Azap, Ö.K.; Arpin, C.; Pascual, A.; Livermore, D.M.; et al. A multinational survey of risk factors for infection with extended-spectrum β-lactamase-producing *Enterobacteriaceae* in non-hospitalized patients. *Clin. Infect. Dis.* **2009**, *49*, 682–690. [CrossRef] [PubMed]

16. Clinical Laboratory Standards Institute (CLSI). Performance Standards for Antimicrobial Susceptibility Testing. In *Proceedings of the Twenty-Sixth Informational Supplement*; CLSI document M100-S26; Clinical Laboratory Standards Institute: Wayne, PA, USA, 2016; Available online: https://clsi.org/ (accessed on 13 March 2017).
17. European Committee on Antimicrobial Susceptibility Testing (EUCAST). Breakpoint Tables for Interpretation of MICs and zone Diameters Version 6.1. 2016. Available online: http://www.eucast.org/fileadmin/src/media/PDFs/EUCAST_files/Breakpoint_tables/v_6.0_Breakpoint_Tables.pdf (accessed on 25 March 2017).
18. Versalovic, J.; Koeuth, T.; Lupski, J.R. Distribution of repetitive DNA sequences in eubacteria and application to fingerprinting of bacterial genomes. *Nucleic Acids Res.* **1991**, *19*, 6823–6831. [CrossRef] [PubMed]

Review

Campylobacter at the Human–Food Interface: The African Perspective

Nikki Asuming-Bediako [1,2], Angela Parry-Hanson Kunadu [3], Sam Abraham [1] and Ihab Habib [1,4,*]

1 School of Veterinary Medicine, College of Science, Health, Education and Engineering, Murdoch University, Perth 6150, Australia; nikkiabed@yahoo.com (N.A.-B.); s.abraham@murdoch.edu.au (S.A.)
2 CSIR-Animal Research Institute, Achimota P.O. Box AH20, Ghana
3 Department of Nutrition and Food Science, University of Ghana, Legon P.O. Box LG134, Ghana; aparry-hanson@ug.edu.gh
4 Veterinary Medicine Department, College of Food and Agriculture, United Arab of Emirates University, Al Ain P.O. Box 1555, UAE
* Correspondence: i.habib@murdoch.edu.au; Tel.: +61-8-9360-2434

Received: 15 April 2019; Accepted: 20 June 2019; Published: 25 June 2019

Abstract: The foodborne pathogen *Campylobacter* is a major cause of human gastroenteritis, accounting for an estimated annual 96 million cases worldwide. Assessment of the true burden of *Campylobacter* in the African context is handicapped by the under-reporting of diarrhoeal incidents and ineffective monitoring and surveillance programmes of foodborne illnesses, as well as the minimal attention given to *Campylobacter* as a causative agent of diarrhoea. The present review of the literature highlights the variability in the reported occurrence of *Campylobacter* in humans and animal food sources across different countries and regions in Africa. *Campylobacter* infection is particularly prevalent in the paediatric population and has been isolated from farm animals, particularly poultry, and foods of animal origin. The reported prevalence of *Campylobacter* in children under the age of five years ranges from 2% in Sudan to 21% in South Africa. In poultry, the prevalence ranges from 14.4% in Ghana to 96% in Algeria. This review also highlights the alarming trend of increased *Campylobacter* resistance to clinically important antimicrobials, such as ciprofloxacin and erythromycin, in humans and food animals in Africa. This review adds to our understanding of the global epidemiology of *Campylobacter* at the human–food animal interface, with an emphasis from the African perspective. Interinstitutional and intersectoral collaborations, as well as the adoption of the One Health approach, would be useful in bridging the gaps in the epidemiological knowledge of *Campylobacter* in Africa.

Keywords: campylobacteriosis; developing countries; one health; zoonoses; antimicrobial resistance

1. Introduction

Campylobacter is a gram-negative, non-spore forming, curved or spiral bacilli, which are oxygen sensitive and prefer to grow under micro-aerobic conditions [1–3]. Some *Campylobacter* species are thermotolerant; for instance, *Campylobacter jejuni* (*C. jejuni*) and *Campylobacter coli* (*C. coli*), which are of critical importance to food safety, grow optimally at 42 °C [2]. In humans, *C. jejuni* and *C. coli* are the main culprits of campylobacteriosis, a very widely recognised enteric illness that can be transmitted to humans through the consumption of undercooked meat, especially poultry, contaminated water and milk, and contact with farm animals such as poultry and livestock [4–7]. It has been widely accepted that improper handling and consumption of contaminated food (notably poultry meat) accounts for the majority of human cases [8,9]. Campylobacteriosis in humans is characterised by watery and/or bloody diarrhoea, abdominal pain, cramps, fever, malaise, and vomiting [10–12]. This is especially dangerous for young children who are more prone to dehydration and loss of nutrients, such as sodium and protein, as a consequence of the diarrhoeal illness [13].

The pathogenesis of *Campylobacter* infection is hypothesised to several mechanisms, however it is not yet fully understood. It has been shown that the expression of genes involved in motility, colonization, epithelial cell invasion, and toxin production play an important role in the disease development [14,15]. Several genes (i.e., *fla*A and *flh*A), are essential for the mobility/passage of *Campylobacter* through the stomach and gut environment [16]. In addition, several proteins (encoded by the *cad*F, *doc*A, *rac*R, *vir*B11, *cia*B, and *iam* genes) on the surface of *Campylobacters* have been shown to promote the adherence and invasion of epithelial cells of the intestine [15,17,18]. *Campylobacter has* also been found to excrete several cytotoxins (encoded by the *cdt*A, *cdt*B, *cdt*C, and *wla*N genes) that contribute to the development of human illness [19,20]. Moreover, *C. jejuni* is able to produce superoxide, dismutase enzyme (encoded by the *sod*B marker), which catalyses the breakdown of superoxide radicals and hence play a major role in defending *Campylobacters* from oxidative damage [21].

In low and middle-income countries, the true incidence of *Campylobacter* is difficult to determine since there are limited systematic surveillance efforts to detect outbreaks and provide isolates that could be used for source-attribution and risk assessment [22]. The infection with *Campylobacter* is regarded as hyper-endemic in many developing countries, due to poor food and environmental sanitation, and close contact with animals at domestic settings in rural and agricultural communities, among many other factors [23]. *Campylobacter* is one of the most frequently isolated bacterial pathogens from the stools of infants with diarrhoea in several developing countries [3,24]. According to the World Health Organisation [9], 40% of the global foodborne disease burden is inflected on children under the age of five years, with the highest burden per population observed in Africa.

Although antimicrobial therapy is not generally indicated in most campylobacteriosis cases, treatment can decrease the duration and reduces the symptoms if it is initiated early in severe cases that warrant antimicrobial intervention [25]. Macrolides (specifically erythromycin) and fluoroquinolones (specifically ciprofloxacin) are considered as the first- and second-choice of antimicrobials, respectively, for the treatment of severe human *Campylobacter* infections [9]. It has been claimed that the spread of antibiotic resistant bacteria/genes to humans through the food chain could be promoted by the uncontrolled extensive use of antibiotics for prophylaxis and treatment in the primary animal production [26]. The situation of antimicrobial resistance in *Campylobacter* is not fully understood across the African continent, despite some reports indicating varying trends at the human-food animal interface [5,25,27–29].

This review collates the knowledge on the epidemiology of *Campylobacter* in humans and food animals in Africa, antimicrobial resistance patterns, and suggestions for management. Specifically, this review aims at elucidating—(i) the prevalence of *Campylobacter* in humans, particularly children, across different regions of the African continent; (ii) the prevalence of *Campylobacter* in foods of animal origin; and (iii) *Campylobacter*'s resistance to antimicrobial agents, notably macrolides and fluoroquinolones.

2. Prevalence of *Campylobacter* in Humans

To facilitate data consolidation and regional comparisons, in this review we collate evidence from literatures based on categorisation of the African continent into five geographical sub-regions (East Africa, Central Africa, West Africa, Southern Africa, and North Africa), in accordance with the United Nations Geoscheme for Africa [30]. Table 1 provides a summary of the published research depicting *Campylobacter* prevalence rates in humans.

Table 1. Prevalence Rates of *Campylobacter* in Humans in Some African Countries.

Region/Country	Population	Sample Size	Prevalence [%]	Genus/Species	Detection Procedure	Reference
EAST AFRICA						
Ethiopia						
Jimma/South Western	Diarrhoeal children under 5 years	227	16.7	*Campylobacter*	Cultural	[5]
			71.1	*C. jejuni*		
			21.1	*C. coli*		
			7.9	*C. lari*		
Gondar/North Western	Diarrhoeal children under age 5	285	15.4	*Campylobacter*	Cultural	[31]
Kola Diba/North Western	Children under age 15	153	10.5	*Campylobacter*	Cultural	[32]
Kenya/South Western	Diarrhoeal children	156	5.8	*Campylobacter*	Cultural	[33]
	Controls	156	1%			
Madagascar	Diarrhoeal children 0–60 months old	2196	9.7	*Campylobacter*	Cultural	[34]
Malawi/Balantyre/Southern	Diarrhoeal children	1941	21	*Campylobacter*	PCR	[35]
	Non-diarrhoeal Children	507	14.1	*Campylobacter*		
Mozambique	Diarrhoeal children	529	1.7	*Campylobacter*	Cultural	[36]
Tanzania						
Morogoro/Eastern	Patients	1195	11.5	*Campylobacter*	Cultural, MALDI-TOF	[37]
			84.1	*C. jejuni*		
			15.9	*C. coli*		
Mwanza/Northern	Diarrhoeal children	300	9.7	*Campylobacter*	Cultural	[38]
Morogoro/Eastern	Diarrhoeal patients	632	9.3	*Campylobacter*	Cultural with Skirrow's protocol and PCR	[39]
			96.6	*C. jejuni*		
			3.4	*C. coli*		

Table 1. *Cont.*

Region/Country	Population	Sample Size	Prevalence [%]	Genus/Species	Detection Procedure	Reference
Uganda, Kampala	Diarrhoeal children	226	9.3	*Campylobacter*	Cultural	[27]
			80.9	*C. jejuni*		
			9.5	*C. lari*		
			4.5	*C. coli*		
			4.5	*C. lari/C. jejuni*		
CENTRAL AFRICA						
Angola, Luanda	Diarrhoeal children under 5	194	15	*Campylobacter*	Multiplex PCR	[40]
		98	23			
	Non diarrhoeal under 5	96	6			
WEST AFRICA						
Burkina Faso, Ouagadougou	Enteritis patients	1246	2.3	*Campylobacter*	Cultural	[41]
			51.8	*C. jejuni*		
			13.8	*C. coli*		
			3.5	*C. upsaliensis*		
Ghana Kumasi	Diarrhoeal and urinary tract infection patients	202	17.3	*Campylobacter*	Cultural	[42]
			40	*C. jejuni*		
			2.8	*C. jejuni subs doylei*		
			37	*C. coli*		
			20	*C.lari*		
Liberia Urban coastal Rural forest	Children 6–59 months	859		*Campylobacter*	Cultural with Skirrow's protocol	[43]
		341 urban	44.9			
		518 rural	28			
Nigeria						

Table 1. *Cont.*

Region/Country	Population	Sample Size	Prevalence [%]	Genus/Species	Detection Procedure	Reference
Sokoto/North Western	Diarrhoeal patients	292	55	*Campylobacter*	Cultural	[44]
Sokoto/North Western	Pregnant women	23	70	*Campylobacter*	Cultural	[45]
Enugu/South Eastern	Diarrhoeal children	514	8.3	*Campylobacter*	Cultural	[46]
			93	*C. jejuni*		
Ilorin/Middle Belt	Diarrhoeal children	306	8.2	*Campylobacter*	Cultural with Butzler type media	[23]
			56	*C. jejuni*		
			44	*C. coli*		
SOUTHERN AFRICA						
Venda/Northern	Human stools	322	10.2	*C. jejuni*	PCR	[47]
			6.5	*C. coli*		
			3.1	*C.conscisus*		
Vhembe/North most	Diarrhoeal stools	565	20.3	*Campylobacter*	Cultural with Cape Town Protocol, PCR	[48]
Cape town/ Coastal	Diarrhoeal stools	5443	40	*C. jejuni*	Cultural with Cape Town Protocol	[49]
			24.6	*C. concius*		
			23.6	*C. upsaliensis*		
Limpopo, North Eastern	HIV individuals	60	20	*Campylobacter*	Cultural	[50]
Durban/South Eastern	Diarrhoeal children under 5	126	21	*Campylobacter*	Cultural	[51]
NORTH AFRICA						
Egypt						
Assiut/South of Cairo	Human	80	27.5	*Campylobacter*	Cultural and Molecular	[52]
Abu Homos/ Northern	Children	6562	9.37	*Campylobacter*	Cultural	[53]

Table 1. *Cont.*

Region/Country	Population	Sample Size	Prevalence [%]	Genus/Species	Detection Procedure	Reference
North of Cairo	Rural children	106	12.3	*C. jejuni*	Cultural	[54]
			2.8	*C. coli*		
South, South East and North of Cairo	Occupational workers	274	8.4%	*Campylobacter*	Biochemical and Molecular	[55]
Zagazig/East Nile Delta	Human	110	2.7	*Campylobacter*	Molecular	[56]
			5.2	*C. jejuni*		
			3.2	*C. coli*		
Giza/Central	Human stools	48	16.66	*Campylobacter*	Cultural	[57]
Abu Homos/Northern	Diarrhoeal children under three years	396	10.5	*Campylobacter*	Cultural	[58]
Cairo	Human	869	16.8	*Campylobacter*	Cultural	[59]
Sudan (Khartoum)	Diarrhoeal children	437	2	*Campylobacter*	Cultural, PCR	[60]

2.1. East Africa

In East Africa, *Campylobacter* infections have been recorded in both rural and urban areas, particularly among children. The prevalence varies between countries (see Table 1), with the highest reported rate being 21% in diarrhoeal children in Malawi [61]. A cross-sectional study conducted from July to October 2012 in the south-western town of Jimma, Ethiopia, detected the presence of *Campylobacter* in the stools of 16.7% of 227 diarrhoeal children under the age of five years [5]. Another study conducted between October 2011 and March 2012 found a prevalence of 15.4% in 285 diarrhoeal children undergoing treatment at the University of Gondar Hospital in northwest Ethiopia [31]. Both studies found that the frequency of *Campylobacter* was higher in malnourished children and in those from households that lacked a source of clean water and had direct contact with domestic animals, particularly hens. Interestingly, the education level of the caregiver, family size, or handwashing before preparing food or eating or after defecation showed no statistically significant association with a positive culture of *Campylobacter*.

In a case-control study conducted in Kisii in south-western Kenya, analysis of 312 stool samples (156 cases and 156 controls) identified the presence of *Campylobacter* in 5.8% and 1% of cases and controls, respectively [33]. In Madagascar, 9.7% of 2196 diarrhoeal stool samples collected from 14 districts during the 2008–2009 rainy season contained *Campylobacter* spp. [34]. In the southern Malawian city of Blantyre, 1941 faecal samples were collected between 1997 and 2007 from children hospitalised with diarrhoea—analysis of these samples indicated that 21% of the samples contained *Campylobacter* compared with 14% of the samples from 507 non-diarrhoeic children [35]. An analysis of 529 stool samples collected from diarrhoeal children at the Manhiça District Hospital in southern Mozambique showed a low *Campylobacter* presence of 1.7% [36].

In Morogoro in eastern Tanzania, Komba, Mdegela, Msoffe, Nielsen and Ingmer [37] detected *Campylobacter* in 11.4% of the stool samples taken from 1195 individuals. The prevalence among symptomatic and young individuals was higher than in asymptomatic and adult individuals. In the northern Tanzanian city of Mwanza, a cross-sectional study of 300 children with acute watery diarrhoea in two hospitals revealed that 9.7% of the stool samples tested positive for *Campylobacter* [38]. In another cross-sectional study in rural and urban areas of Morogoro in eastern Tanzania from January 2003 to December 2004, the prevalence of *Campylobacter* was reported as 9.3% in 632 human stool samples, with *C. jejuni* accounting for more than 90% of the positive isolates [39].

Similar to the studies from Tanzania, Mshana, Joloba, Kakooza and Kaddu-Mulindwa [27] recorded a 9.3% isolation rate of *Campylobacter* in 226 stool samples from diarrhoeal children attending the Mulago Hospital in Kampala, the capital city of Uganda, with *C. jejuni* being the most frequently detected species (80.9%). Mshana et al. further reported a higher infection rate of 10.9% in children under two years of age compared with a rate of 8.5% in children over two years of age. The researchers proposed that the increase in protective antibodies with age may account for the decreased rate of infection in older children.

2.2. Central Africa

There are limited published studies from the central African region. An analysis of 194 stool samples from 98 children with acute diarrhoea and 96 children without diarrhoea under the age of five in Angola's capital city of Luanda found that *Campylobacter* was present in 15% of the samples overall, with 23% present in the stools of diarrhoeic children compared with 6% in the stools of non-diarrhoeic children [40]. Multiplex real-time polymerase chain reaction (mPCR) was used to analyse the samples. Other pathogens, including *Escherichia coli*, *Salmonella*, *Cryptosporidium*, and *Shigella*, were detected in all the samples, regardless of diarrhoeal status.

2.3. West Africa

In Ouagadougou, the capital of Burkina Faso, Sangaré, Nikiéma, Zimmermann, Sanou, Congo-Ouédraogo, Diabaté, Diandé and Guissou [41] collected stool samples from 1246 enteritis patients from 2006 to 2008 and reported a *Campylobacter* isolation rate of 2.3%, with *C. jejuni* accounting for 51.8%, *C. coli* accounting for 13.8%, and *C. upsaliensis* accounting for 3.5%. In Kumasi, the capital of Ghana, Karikari, Obiri-Danso, Frimpong and Krogfelt [42] reported a *Campylobacter* prevalence of 17.3% in 202 patients who visited the Komfo Anokye Teaching Hospital from May to August 2013.

In Liberia, researchers reported a noteworthy *Campylobacter* isolation rate of 44.9% in 341 children from a crowded urban slum compared with only 28% in 518 children from a cleaner rural area. [43]. The children were aged between six months and five years. The authors found that the prevalence of *Campylobacter* increased in children aged over 18 months, arguing that this may be attributed to increased contact with animals and the environment. Other risk factors for infection were the post-weaning consumption of contaminated food and water.

In Nigeria, Africa's most populous country, analysis of 292 stool samples taken from people in hospitals across four agricultural zones in the north-western state of Sokoto detected the presence of *Campylobacter* in 55% of the samples [44]. Another study in the same state found that 70% of 23 pregnant women and 43% of 57 non-pregnant women were stool-positive for *Campylobacter* [45]. In this study, exposure was attributed to poor environmental conditions in the homes of patients [45]. In Enugu, in the south-eastern state of Nigeria, a lower isolation rate of 8.3% was found in 514 children under the age of five years, with *C. jejuni* accounting for 93% of the positive isolates [46]. Similarly, Samuel, Aboderin, Akanbi II, Adegboro, Smith and Coker [23] detected *Campylobacter* in 8.2% of the stool samples from 306 diarrhoeal children in Ilorin in the middle belt zone of Nigeria, with all positive isolates being found in children under the age of two years. The authors indicated that the key risk factors were exposure to an unclean environment and the consumption of contaminated foods and water after weaning [23].

2.4. Southern Africa

In the Venda region, located in the northern part of South Africa, *C. jejuni* was detected in 10.2% of 322 stool samples collected from patients admitted to a hospital [47]. Although other pathogens such as *H. pylori*, *Arcobacter butzleri*, *A. skirowii*, and *A. cryaerophilus* were also present, *C. jejuni* was significantly associated with diarrhoea. In another study in South Africa, samples of diarrhoeal stools were taken from 565 people in rural areas in the northernmost district of Vhembe, and analysis using the filtration method detected *Campylobacter* in 20.3% of the samples. A higher isolation rate of 30.4% was seen in the samples taken from children under the age of two [48]. Lastovica [49] analysed 5443 diarrhoeal stool samples collected from the Red Cross Children's Hospital in the coastal city of Cape Town between 1990 and 2005 and found that 40% contained *C. jejuni*, 24.6% contained *C. concisus*, and 23.6% contained *C. upsaliensis*. The isolation was carried out using the Cape Town Protocol, which may have contributed to the high levels detected. In the rural Limpopo province of north-eastern South Africa, *Campylobacter* was isolated from 20% of the stool samples taken from 60 HIV-positive individuals with chronic diarrhoea [50]. Forty of the 60 individuals tested were positive for other diarrhoeal agents, including *E. coli*, *Shigella*, *Salmonella*, *Plesiomonas shigelloides*, and *Aeromonas* spp. HIV infection is known to seriously compromise immunity; hence, patients were susceptible to a wide range of infections. Further, in two interrelated studies undertaken in a Durban hospital, *Campylobacter* was found in 21% of the stool samples taken from 126 malnourished inpatient children compared with 7% of the stool samples taken from 352 randomly selected outpatient children [51]. Other pathogens such as *Salmonella*, *E. coli*, and *Shigella* were isolated from both groups. Malnutrition compromises the body's defence system, thus increasing susceptibility to infection.

2.5. North Africa

A study by Abushahba [52] in Assiut in Egypt, located about 375 km south of the capital Cairo, found that 27.5% of 80 human stool samples screened positive for *Campylobacter*. Of the participants in that study, 33 were infants under the age of 12 months. The key risk factors for infection were impaired immunity and residential conditions in villages, with poor hygiene and poultry rearing in households. In a prospective study conducted in Abu Homos, an agricultural community in northern Egypt, from 1995 to 2003, *Campylobacter* was isolated from 9.37% of 6562 faecal samples collected from 1057 children [53]. In the Gharbia Governorate located in the Nile Delta region of Egypt, El-Tras, Holt, Tayel and El-Kady [54] found a prevalence of 12.3% and 2.8% for *C. jejuni* and *C. coli*, respectively, in 106 rural children from households that owned poultry. Poultry is a major reservoir for *Campylobacter* and is, therefore, an important source of transmission—backyards or coops with wet litter and poor sanitation increase the risk of human exposure. In Zagazig in the eastern part of the Nile Delta, Awadallah, Ahmed, El-Gedawy and Saad [56] detected a *Campylobacter* prevalence of 2.7% in 110 stool samples sourced from the El-Ahrar General Hospital from September 2012 to April 2014. Hassanain [57] reported a higher *Campylobacter* prevalence of 16.66% in 48 human faecal samples collected from individuals in contact with food-producing animals in the Giza Governorate in central Egypt.

From 1995 to 1998, a case-control study of 397 children under the age of three years in the Abu Homos region of northern Egypt reported 3477 episodes of diarrhoea of which 366 (10.5%) were associated with *Campylobacter* [58]. The presence of animals in the house, particularly in cooking areas, and other unhygienic conditions were major risk factors for infection. Breastfeeding did not appear to reduce the risk of *Campylobacter*-associated diarrhoea, but a reduced risk was associated with adequate toilet facilities. At Abbassia Fever Hospital in Cairo, 869 *Salmonella*, *Shigella*, and *Campylobacter* strains were isolated from 6278 patients who visited the hospital from January 1986 to December 1993. Although *Salmonella* was the predominant strain at 53.5%, *Campylobacter* showed a prevalence of 16.8%, with *C. jejuni* present in 92 of the 146 *Campylobacter*-positive isolates [59].

In Sudan, an isolation rate of 2% was found in 437 stool samples of diarrhoeal children collected from January to December 2013 in suburban Khartoum [60]. The samples were also colonised by other pathogens, including *E. coli*, rotavirus A, *Shigella*, *Salmonella,* and *Giardia intestinalis*. Bacterial agents were the most common cause of diarrhoea and children over two years of age were frequently affected. Contaminated hands are a common source of foodborne infections and, given that proper handwashing is a challenge for this age group, this may have been a reason for the increased prevalence of diarrhoea [60].

3. Prevalence of *Campylobacter* in Foods of Animal Origin

Meat, eggs, milk, and other products from animals represent an important part of the diet of Africans [62]. On the African continent, varying rates of *Campylobacter* prevalence in food of animal origin have been reported ranging from 2% in beef to 90% in chicken carcasses. Table 2 provides a summary of the *Campylobacter* prevalence in foods of animal origin.

Table 2. Occurrence of *Campylobacter* in Foods of Animal Origin.

Region/Country	Product	Sample Size	Percentage Positive	Genus/Species	Detection Procedure	Reference
EAST AFRICA						
Ethiopia/South East Addis Ababa	Sheep and goat carcass	398	10.1	*Campylobacter*	Cultural	[63]
			72.5	*C. jejuni*		
			27.5	*C.coli*		
Kenya/Nairobi	Chicken	100	77	*Campylobacter*	Cultural	[64]
	Beef	50	2			
Tanzania						
Northern, South Western, Eastern	Beef carcass	253	9.5	*Campylobacter*	mPCR	[65]
	Raw milk	284	13.4			
Morogoro/Eastern	Duck intestines	90	80	*Campylobacter*	Cultural with Skirrow's protocol	[66]
Morogo/Eastern	Cattle carcass	107	9.3	*Campylobacter*	Cultural with Skirrow protocol	[67]
CENTRAL AFRICA						
Cameroon/Yaounde	Chicken	150	90	*Campylobacter*	Cultural	[68]
Congo DR/Lubumbashi	Goat meat	177	41.2	*Campylobacter*	PCR	[69]
	Goat stomach	86	37.2			
	Ready to eat goat skewer	139	23.7			
WEST AFRICA						
Burkina Faso/Ouagadougou	Chicken carcass	20	50	*Campylobacter*	Cultural	[70]
Ghana						

Table 2. *Cont.*

Region/Country	Product	Sample Size	Percentage Positive	Genus/Species	Detection Procedure	Reference
Kumasi	Poultry carcass	132	21.9	*Campylobacter*	Cultural	[71]
			79	*C. jejuni*		
			14	*C. coli*		
			4	*C. jejuni subs doylei*		
			3	*C. lari*		
Kumasi	Cattle carcass	110	34.5	*Campylobacter*	Cultural, mPCR	[72]
	Goat carcass	134	23.9			
	Sheep carcass	117	35.9			
	Pig carcass	102	36.3			
Nigeria						
Sokoto/North Western	Raw milk	146	4.8	*Campylobacter*	Cultural	[73]
			100	*C. jejuni*		
Sokoto/North western	Chicken	681	81.9	*Campylobacter*	Cultural	[74]
			60.9	*C. jejuni*		
			39.1	*C.coli*		
Senegal/Dakar	Chicken	300	56	*Campylobacter*	Cultural	[75]
SOUTHERN AFRICA						
South Africa Gauteng/North	Chicken carcass	99	32.3	*Campylobacter*	Cultural	[76]
NORTH AFRICA						
Algeria						
Middle area	Turkey neck skin	100	55	*Campylobacter*	Cultural	[77]

Table 2. *Cont.*

Region/Country	Product	Sample Size	Percentage Positive	Genus/Species	Detection Procedure	Reference
Middle area; Algiers, Bouira, Boumerdes	Chicken neck, giblets	346	17.9	*Campylobacter*	Cultural with Butzler medium and Skirrow's protocol	[78]
Egypt						
Assiut/South of Cairo	Chicken	104	24	*Campylobacter*	Cultural	[52]
Qena city/Southern	Milk, cheese, yogurt	150	24.6	*Campylobacter*	Cultural, PCR	[79]
Cairo, Minya, Qalubiya and Fayoum	Yogurt	344	1.2	*Campylobacter*	Cultural	[55]
	Raw milk	457	2.0			
	Cheese	288	1.7			
	Chicken intestine	211	12.8			
	Chicken meat	9.6	680			
Zagazig/Eastern Nile Delta	Chicken breast	64	47.5	*Campylobacter*	Cultural	[56]
	Chicken thighs	64	25.9			
	Chicken skin	64	21.6			
Abou Homos/Northern	Milk, milk products	227	2.64	*C. jejuni*	mPCR	[80]
Morocco/ Oujda/Eastern	Poultry carcass	50	62	*Campylobacter*	Cultural, Hippurate hydrolysis	[81]
			90	*C. jejuni*		
			10	*C. coli*		

3.1. East Africa

Sheep and goat meat are an important part of the Ethiopian diet. A cross-sectional study of 398 sheep and goat carcasses from a private export abattoir in Debre-Zeit, 45 km south-east of Addis Ababa, between October 2007 and March 2008 found that *Campylobacter* was present in 10.1% of the samples, with *C. jejuni* accounting for 72% of the isolates [63]. The highest bacterial isolation was from the breast region of the carcasses, resulting from cross-contamination from intestinal contents during manual skinning, evisceration, processing, and contact with processors' hands or knives. Additionally, washing of carcasses carries microorganisms from other parts of the body to the breast region. Therefore, effective hygienic practices and attentive evisceration during slaughter and dressing are important measures to reduce contamination.

In Kenya, Osano and Arimi [64] analysed 100 chicken and 50 beef samples from butchers, supermarkets, and markets in Nairobi and reported the presence of *Campylobacter* in 77% and 2% of chicken and beef samples, respectively. *C. jejuni* was the dominant species isolated from chicken, emphasising its potential role in zoonotic transmission between humans and poultry. In Tanzania, Kashoma, Kassem, John, Kessy, Gebreyes, Kazwala and Rajashekara [65] reported the presence of *Campylobacter* in 9.5% of 253 beef carcasses and 13.4% of 284 unpasteurised raw milk samples using mPCR. This was conducted from April 2013 to March 2014 and samples were sourced from Arusha, Iringa and Morogoro in Tanzania. The milk was obtained from milk vendors and milk tanks at milk collecting centres, while the beef swabs were taken from dressed carcasses. This study illustrates that the consumption of raw milk is a route for the transmission of *Campylobacter*. Cattle also carry *C. jejuni* and cross-contamination could occur during slaughter and milking. Therefore, slaughter and milking should be carried out using hygienic methods to minimise the transmission of pathogens to meat and milk and, consequently, to humans. Similarly, *Campylobacter* was detected in 9.3% of 107 cattle carcasses sampled from an abattoir in Morogoro in east Tanzania, while in meat shops, it was detected in 1.7% of samples [67]. The authors reported that the dressing and sale of meat was carried out in unhygienic environments and some shops sold chicken in addition to beef, thereby enhancing the risk of transfer.

3.2. Central Africa

In Cameroon, Nzouankeu, Ngandjio, Ejenguele, Njine and Wouafo [68] reported that 90% of 150 retail chickens obtained from eight markets in the capital Yaounde from February 2006 to January 2007 contained *Campylobacter* spp.. *E. coli* and *Salmonella* were also isolated from the samples using culture-based methods. *Campylobacter* is commensal in poultry, its primary host, and the risk of cross-contamination to the carcass during slaughter and processing is high if this is not carried out carefully and hygienically. Nzouankeu et al. further suggested the need to monitor poultry for pathogens and to minimise cross-contamination.

In a study of retail goat meat outlets in Lubumbashi in the Democratic Republic of the Congo, *Campylobacter* was found in 41.2% of 177 goat meat samples, 37.2% of 86 goat stomach samples, and 23.7% of 139 ready-to-eat goat skewers using polymerase chain reaction (PCR) [69]. The outlets comprised open air and semi-open-air markets, snack bars, and bars. The authors noted that the slaughter of animals was undertaken in unhygienic facilities and retail points, and no efforts were made to ward off flies or other vermin. However, cooking decreased the prevalence of *Campylobacter*, as evidenced by the lower rate recorded in the ready-to-eat goat skewers sourced from the same outlets. Thorough cooking of meat is an effective way of reducing the risk of infection in humans.

3.3. West Africa

In Ouagadougou, Burkina Faso, *Campylobacter* was detected in 50% of 20 poultry carcasses sourced from retail markets [70]. The carcasses were sold on tables in ambient temperatures without protection from dust or flies, the vendors did not wear gloves or aprons, and the retail environment was infested with lizards, rodents, and avian species. Kagambèga, Thibodeau, Trinetta, Soro, Sama, Bako, Bouda,

Wereme N'Diaye, Fravalo and Barro [70] suggested that, although poultry is widely consumed in Burkina Faso, patients with diarrhoea are not routinely sampled for *Campylobacter*. However, in neighbouring Ghana, a lower isolation rate of 21.9% was detected in 132 poultry carcasses randomly selected from the Kejetia poultry slaughter unit in Kumasi [71]. *Campylobacter* is known to colonise the intestinal tract of poultry and cross-contamination can occur during slaughtering and processing if the carcass is not properly handled. Karikari, Obiri-Danso, Frimpong and Krogfelt [72] found that *Campylobacter* were present in 34.5% of beef, 23.9% of goat, 35.9% of sheep, and 36.3% of pig carcasses sampled from the Kumasi abattoir. The contamination of carcasses during manual skinning, evisceration, and processing at the abattoir were the reasons attributed to the prevalence rates recorded.

In Sokoto in north-western Nigeria, Salihu, Junaidu, Magaji and Rabiu [73] detected the presence of *Campylobacte*r in 4.8% of 146 raw milk samples from lactating herds between October 2007 and September 2008. All the positive samples contained *C. jejuni* biotype I. The authors suggested that animal and animal products were reservoirs for human infections. Analysis of breed distribution showed that White Fulani breeds had a higher prevalence (5.4%) than Sokoto Gudali breeds (4.7%) or Friesian–Sokoto Gudali crossbreeds (0.0%). Using culture methods, Salihu, Junaidu, Magaji, Abubakar, Adamu and Yakubu [74] detected *Campylobacter* in 81.9% of 681 chicken samples in Sokoto from November 2007 to October 2008. *C. jejuni* accounted for 60.9% of the isolates, followed by *C. coli* at 28%, and *C. lari* at 7%. Biotyping showed a prevalence of biotype I in *C. jejuni, C. coli,* and *C. lari. C. jejuni* I and *C. coli* I are found in humans; hence, the results were indicative of the zoonotic nature of the pathogen.

From January 2001 to May 2002, 300 fresh, refrigerated, or frozen chicken carcasses from retail outlets in Dakar in Senegal were analysed, with *Campylobacter* being present in 56% of the samples [75]. The contamination rates were found to be highest in the fresh samples at 76% and lowest in the frozen samples at 28%. Fifty-three percent of the refrigerated samples were also contaminated. *Campylobacter* is stress sensitive and freezing affects its viability, hence the lower rates found in frozen samples. Therefore, hygienic handling practices during slaughter and processing, and adequate personal hygiene and cooking techniques are important for infection control.

3.4. Southern Africa

In South Africa, van Nierop, Duse, Marais, Aithma, Thothobolo, Kassel, Stewart, Potgieter, Fernandes, Galpin and Bloomfield [76] found that *Campylobacter* was present in 32.3% of 99 fresh and frozen chicken carcasses sourced from retailers in Gauteng in the country's north. Carcasses from supermarkets were more frequently contaminated with *Campylobacter* than those from butchers, which were more frequently contaminated with *Salmonella*. Similarly, in a study in Senegal, *Campylobacter* was isolated more frequently from fresh chicken than from frozen chicken. Culture methods detected an isolation rate of 32.3%, whereas PCR detected a rate of 43.4%. Detection methods play a key role in detecting prevalence, with PCR reported to be more sensitive in the detection of pathogens. However, the authors cautioned that the use of PCR may have resulted in false positives because of DNA contamination, the presence of inhibitory substances in the enrichment broths or the identification of non-viable pathogens as viable.

3.5. North Africa

In Algeria, Bouhamed, Bouayad, Messad, Zenia, Naim and Hamdi [77] found that *Campylobacter* was present in 55% of 100 turkey neck skins sampled from three traditional and one modern abattoir located in the country's middle regions, Algiers, Bouïra, and Boumèrdes. The abattoirs were characterised by a lack of disinfection protocols and sterilisation of equipment, and by the use of dirty uniforms. These are important factors that enhance the spread of pathogens. However, Laidouci, Mouffok and Hellal [78] found a lower presence of *Campylobacter* (17.9%) in 346 chicken neck and giblets sourced from Algeria.

In Egypt, 24% of 104 chicken carcasses from two slaughterhouses in the Assiut Governorate south of Cairo contained *Campylobacter* [52]. An analysis of 150 raw milk, kareish cheese, and yoghurt samples from September 2014 to February 2015 found that 24.6% tested positive for *Campylobacter* spp. [79]. All the samples tested positive for *C. jejuni* and negative for *C. coli*. The samples were obtained from the local market and street vendors in the city of Qena in southern Egypt. After the samples were cultured, mPCR was used to confirm the isolates. The presence of *Campylobacter* in raw milk was attributed to contamination during on-farm milking processes or to poor post-milking storage and handling conditions. Unhygienic methods used in the preparation, processing, handling, and sales of kareish cheese accounted for it having the highest presence of *Campylobacter* among the products tested. In a study conducted across the governorates of Cairo, Minya, Qalyubia, and Fayoum in Egypt, Omara, Fadaly and Barakat [55] found that *Campylobacter* was present in 12.8% of raw chicken intestines, 9.6% of 680 raw chickens, 1.2% of 344 yoghurt samples, 2.0% of 457 raw milk samples, and 1.7% of 288 kareish cheese samples. The contamination was attributed to unsanitary food production and storage practices. Further, Awadallah, Ahmed, El-Gedawy and Saad [56] found that *Campylobacter* was present in 47.5%, 25.9%, and 21.6% of chicken skins, chicken thighs, and chicken breasts, respectively. These products were sourced from a slaughterhouse in Zagazig, a city in the eastern part of the Nile Delta—the authors found that refrigeration and particularly freezing reduced the counts of viable *Campylobacter* cells. In contrast, in Abu Homos in northern Egypt, a study detected *Campylobacter* in only 2.64% of 227 milk and milk product samples [80]. Using culture methods and mPCR assays, raw milk and fresh domiati cheese (a moderately slated enzyme-coagulated soft cheese) samples tested positive for *Campylobacter*. The remaining milk products, laban rayeb (traditional low-fat fermented milk), stored domiati cheese (a highly salted enzyme-coagulated soft cheese), zabady (Egyptian yoghurt), ras cheese (a hard cheese variety), and kareish cheese (an acid-coagulated soft cheese) were all negative for *Campylobacter*. The products that tested positive were contaminated with *C. jejuni*, which had survived preservation methods better than expected, implying that the pathogen has the ability to develop adaptive strategies to aid survival in food preservation conditions.

In Morocco, *Campylobacter* was detected in 62% of 50 retail poultry samples sourced from the Oujda area in the eastern part of the country [81]. Similar to other countries, the high level of contamination was attributed to poultry being the primary host of *Campylobacter* and cross-contamination occurring at unhygienic slaughter and retail points. The authors further identified the antibacterial effects of common condiments, finding that that 1% v/v lemon juice and vinegar and 2% v/v cinnamon and sodium chloride had high inhibitory effects. However, onion, ginger, black pepper, cumin, parsley, garlic, and saffron had minimal or no effect.

4. Antimicrobial Resistance (AMR) Patterns of *Campylobacter* in Africa

According to the World Health Organisation [22], surveillance of AMR in *Campylobacter* has identified important levels of resistance to erythromycin and fluoroquinolones in many parts of the world, which appears to be associated with the use of these drugs in poultry and livestock production systems. Some epidemiological studies in humans and animals have established a relationship between antibiotic use and antimicrobial resistance [82–84]. Campylobacteriosis is typically a self-limiting disease, which does not usually require antibiotic treatment; however, in some cases, antibiotics may be administered. Fluoroquinolones and macrolides such as ciprofloxacin and erythromycin, respectively, are recommended for the treatment of *Campylobacter* infections in humans [24,85]. However, given their abuse and misuse, resistance to these drugs has emerged. Resistance to macrolides such as erythromycin and azithromycin is primarily attributed to mutations in the m23SRNA gene and decreased permeability of *Campylobacter* cell walls, while resistance to fluoroquinolone is typically moderated by mutation of DNA gyrase and efflux through the outer and inner membranes [86].

In this review we will focus on the resistance patterns of *Campylobacter* spp. from African environments especially to fluoroquinolones and macrolides as these antimicrobials are the most commonly used for the treatment of *Campylobacter* infections in humans. Table 3 provides a summary of resistance trends.

Table 3. Antibiotic Resistance Trends of *Campylobacter* in Some African Countries.

Region/Country	Source	Species	Antibiotic	Resistance %	Method Used	Reference
EAST AFRICA						
Ethiopia/South Western	Humans, diarrhoeal children	*Campylobacter*	Tetracycline	39.5	Kirby-Bauer Disk Diffusion	[5]
			Chloramphenicol	31.6		
Kenya, North eastern Nairobi	Backyard chicken	*C. jejuni*	Tetracycline	71	PCR	[87]
			Ciprofloxacin	71		
			Nalidixic acid	77.4		
			Chloramphenicol	25.8		
Mozambique	Diarrhoeal children	*Campylobacter*	Tetracycline	22	Disk Diffusion	[36]
			Ciprofloxacin	11		
			Chloramphenicol	11		
			Nalidixic acid	11		
Tanzania						
Morogo/Eastern	Free range ducks	*Campylobacter*	Erythromycin	20–50	Disk Diffusion	[66]
Iringa/North, Morogoro/Eastern Arusha/South West	Milk, beef	*Campylobacter*	Ciprofloxacin	9.3	Disk Diffusion	[65]
			Erythromycin	53.7		
			Tetracycline	18.5		
			Azithromycin	42.6		
			Nalidixic acid	64.8		
			Chloramphenicol	13		
			Ciprofloxacin	11.8		
			Erythromycin	70.6	Broth micro-dilution	
			Tetracycline	17.7		

Table 3. *Cont.*

Region/Country	Source	Species	Antibiotic	Resistance %	Method Used	Reference
Morogoro/Eastern	Humans	*Campylobacter*	Ciprofloxacin	22.1	Disk Diffusion	[37]
Uganda/Kampala	Humans	*Campylobacter*	Ciprofloxacin	5	Disk Diffusion	[27]
WEST AFRICA						
Burkina Faso/Ouagadougou	Humans	*Campylobacter*	Ciprofloxacin	13.8	Disk Diffusion	[41]
			Tetracycline	10.3		
			Erythromycin	10.3		
			Nalidixic acid	34.5		
Ivory Coast/Abidjan	Chicken	*C. jejuni*	Ciprofloxacin	38.5	Disk Diffusion	[6]
			Erythromycin	10.3		
			Nalidixic acid	79.5		
		C. coli	Ciprofloxacin	43.2	Disk Diffusion	
			Erythromycin	8.1		
			Nalidixic acid	78.4		
Ghana						
Kumasi/South	Humans	*Campylobacter*	Erythromycin	92.3–100	Disk Diffusion	[42]
			Tetracycline	92.3–93.3		
Kumasi/South	Poultry carcass and faeces	*Campylobacter*	Quinolones	41–86	Kirby-Bauer Disk Diffusion	[71]
			Erythromycin	100		
			Tetracycline	97–100		
Kumasi/South	Livestock	*Campylobacter*	Quinolones	7–69	Kirby-Bauer Disk Diffusion	[72]
			Erythromycin	97–100		
			Tetracycline	48–94		

Table 3. *Cont.*

Region/Country	Source	Species	Antibiotic	Resistance %	Method Used	Reference
Nigeria						
Sokoto/North Western	Poultry cloacal swabs	*Campylobacter*	Erythromycin	11.6	Disk Diffusion	[88]
Osogbo/South Western	Humans	*Campylobacter coli*	Ciprofloxacin	0	Kirby-Bauer Disk Diffusion	[25]
			Nalidixic acid	66		
			Erythromycin	0		
			Tetracycline	0		
Ilorin/Middle belt	Humans	*Campylobacter*	Ciprofloxacin	0	Kirby-Bauer Disk Diffusion	[23]
			Erythromycin	0		
			Nalidixic acid	24		
			Tetracycline	12		
SOUTHERN AFRICA						
Durban/South Eastern	Humans	*Campylobacter*	Tetracycline	64	PCR	[89]
	Chicken			68		
Durban/South Eastern	Humans	*C. jejuni*	Erythromycin	31.5	Broth microdilution	[90]
			Azithromycin	50		
		C. coli	Erythromycin	38.9		
			Azithromycin	77		

Table 3. *Cont.*

Region/Country	Source	Species	Antibiotic	Resistance %	Method Used	Reference
Vende/Northern	Cattle	C. jejuni	Ciprofloxacin	33.3	Disk diffusion	[91]
			Erythromycin	42.9		
			Nalidixic acid	26.2		
			Tetracycline	31		
	Chicken		Ciprofloxacin	29		
			Erythromycin	56.7		
			Nalidixic acid	47.8		
			Tetracycline	33.3		
	Cattle	C. coli	Ciprofloxacin	56.3		
			Erythromycin	6.8		
			Nalidixic acid	37.5		
			Tetracycline	62.5		
	Chicken		Ciprofloxacin	37.5		
			Erythromycin	43.8		
			Nalidixic acid	31.3		
			Tetracycline	43.8		

Table 3. *Cont.*

Region/Country	Source	Species	Antibiotic	Resistance %	Method Used	Reference
Kwazulu Natal	Commercial free range broilers	*Campylobacter*	Tetracycline	100	Agar Dilution	[92]
			Ciprofloxacin	95.4		
			Erythromycin	87.9		
	Rural birds		Tetracycline	21.6		
			Ciprofloxacin	7.9		
			Erythromycin	0		
	Industrial broiler		Tetracycline	98.9		
			Ciprofloxacin	15.9		
			Erythromycin	47.6		
	Industrial layer		Tetracycline	100		
			Ciprofloxacin	17.7		
			Erythromycin	43.7		
Kwazulu Natal	Broiler	*Campylobacter*	Tetracycline	98.2	Agar Dilution	[28]
			Erythromycin	50		
			Ciprofloxacin	8.9		
			Nalidixic acid	35.7		
	Layer		Tetracycline	100		
			Erythromycin	42.9		
			Ciprofloxacin	23.8		
			Nalidixic acid	52.4		

Table 3. *Cont.*

Region/Country	Source	Species	Antibiotic	Resistance %	Method Used	Reference
NORTH AFRICA						
Algeria						
Bouira/Middle area	Turkey carcasses	*Campylobacter*	Tetracycline	81.3	Disk Diffusion	[77]
			Ciprofloxacin	75		
			Erythromycin	25		
			Nalidixic acid	87.5		
Middle area/Algiers, Bouira &, Boumerdes	Sheep, calves broiler carcasses	*Campylobacter*	Ciprofloxacin	91.6	Disk Diffusion	[29]
			Erythromycin	88.5		
			Tetracycline	44.7		
			Nalidixic acid	96.8		
Egypt						
Giza	Poultry	*Campylobacter*	Erythromycin	58.8	Disk Diffusion	[57]
			Tetracycline	58.8		
			Chloramphenicol	64.7		
			Erythromycin	50		
	Human		Tetracycline	75		
			Chloramphenicol	50		
Cairo	Human	C. jejuni	Nalidixic acid	40	Disk diffusion	[59]
			Tetracycline	6		
			Erythromycin	9		
			Chloramphenicol	3		
		C. coli	Nalidixic acid	24		
			Tetracycline	24		
			Erythromycin	10		
			Chloramphenicol	0		

4.1. East Africa

In Jimma, Ethiopia, erythromycin, ciprofloxacin, gentamicin, and nalidixic acid were reportedly effective for more than 80% of *Campylobacter* species isolated from diarrhoeal children using the Kirby–Bauer disc diffusion method. However, 39.5% of the isolates were resistant to tetracycline and 31.6% to chloramphenicol, while resistance to ciprofloxacin and erythromycin was 12.0% [5]. Multidrug resistance was also observed in 78.9% of the isolates. The high resistance to tetracycline, for example, is attributed to its wide availability without prescription; hence, overuse may have resulted in selective pressure on bacteria, thereby making them resistant.

In Kenya, 74% of *C. jejuni* isolates from backyard chickens in Thika, north east of the capital Nairobi, showed resistance to nalidixic acid, while 71% were resistant to each of tetracycline and ciprofloxacin [87]. In that study in Kenya [86], the uncontrolled use and ease of access to antibiotics by small-scale farmers were suggested as reasons for the high resistance rates recorded. In Mozambique, a resistance rate of 11% was reported for ciprofloxacin and nalidixic acid in stool samples of diarrhoeal children at the Manhica Hospital [36]. Such relatively low rates might be expected as the use of fluoroquinolones is not recommended in young children [36].

In Tanzania, *Campylobacter* isolates from milk and beef collected from the Eastern (Morogoro), Northern (Arusha) and South Western (Iringa) parts of the country showed resistance to ciprofloxacin ranged from 9.3% to 11.8%, while the range for erythromycin resistance was between 53.7% and 70.6% [65]. The authors also noted that, in Tanzania, the macrolide, Tylosin, was extensively used for treating respiratory infections like *Mycoplasma* in cattle and its usage for treatment and growth promotion might have contributed to the selection of resistant strains to erythromycin. Furthermore, isolates from free-range ducks in the same city, were susceptible to nitrofurantoin and amikacin, while they showed 74% resistance to tetracycline [66].

4.2. West Africa

In Burkina Faso, isolates from human stools showed 13.8% resistance to ciprofloxacin, 10.3% to tetracycline, and 10.3% to erythromycin [41]. A study in Abidjan, Ivory Coast Goualie, Essoh, Elise Solange, Natalie, Souleymane, Lamine Sebastien and Mireille [6], reported 79.5%, 38.5%, 17.9%, and 10.3% resistance of poultry *C. jejuni* isolates to nalidixic acid, ciprofloxacin, amoxicillin, and erythromycin, respectively. *C. coli*, on the other hand, showed resistance rates of 78.4%, 43.2%, 13.5%, and 8.1% to nalidixic acid, ciprofloxacin, amoxicillin, and erythromycin, respectively. The authors noted that the unrestricted use of antibiotics for treatment and as growth promoters contributed to the resistance.

In neighbouring Ghana, a resistance of 92.3–100% was reported for erythromycin, 92.3–93.3% for tetracycline, and 0% for imipenem for isolates from patients in Kumasi [42]. Erythromycin resistance arises as a result of prolonged exposure and evidently, as noted by the authors, erythromycin had been on the Ghanaian market for a long time and was subjected to abuse and misuse. Additionally, the isolates from carcasses and faeces of poultry in Kumasi were all completely sensitive to imipenem. However, the resistance to quinolones ranged from 41–86%, 100% to erythromycin and 97–100% to tetracycline [71]. Furthermore, the isolates from livestock showed 7–69% resistance to quinolones, 97–100% to erythromycin, and 48–94% to tetracycline. There was 0% resistance to imipenem [72]. In Sokoto, (North Western Nigeria), the isolates from cloacal swabs of poultry showed resistance to most of the antibiotics except chloramphenicol. Multidrug-resistant traits were found in 82.1% of the isolates. Resistance to ciprofloxacin was 19.6%, ampicillin 32.1%, tetracycline 24.1%, and erythromycin 11.6% [88]. Due to the low levels of resistance recorded for erythromycin, it was recommended for treatment. The zero resistance to chloramphenicol was because it had been banned for use in both humans and animals [88]. Additionally, in Ilorin in the middle belt of Nigeria, Samuel, Aboderin, Akanbi II, Adegboro, Smith and Coker [23] reported resistance of human isolates to nalidixic acid (24%) and tetracycline (12%) but the isolates showed no resistance to erythromycin and ciprofloxacin. Besides, the isolates from diarrhoeal stools of children below age 3 years in Osogbo, South Western

Nigeria [25], were reported to be sensitive to erythromycin, tetracycline, and ciprofloxacin, and were hence recommended for treatment of *Campylobacter* infection. The former study pointed out that all isolates were not sensitive to cotrimoxazole [25].

4.3. Southern Africa

A study in Durban, South Africa, reported 68% and 64% resistance to tetracycline in the isolates of *Campylobacter* from chicken and humans, respectively [89]. Also in Durban, Shobo, Bester, Baijnath, Somboro, Peer and Essack [90], reported that 31.5%, 50%, and 25.9% of *C. jejuni* isolates from humans were resistant to erythromycin, azithromycin, and tetracycline, respectively; while 38.9%, 77%, and 55.6% of *C. coli* strains showed resistance to erythromycin, azithromycin, and erythromycin, respectively. In *Campylobac*ter, resistance to tetracycline is primarily mediated by a ribosomal protection protein (*tet*O) that is transferred as a plasmid-encoded gene or in the chromosome where it is not self-mobile [90].

Production systems and bird type also have an influence on isolation rates and resistance. In the Kwazulu Natal province of South Africa, it was established that the isolates from commercially raised, i.e., free range and industrial chickens were highly resistant to tetracycline (98.9–100%). Resistance to gentamicin and streptomycin was 1.6% and 11.5% respectively in the commercial free-range broiler, 1.7% and 16.4% in industrially raised broilers, and 12.9 and 40% respectively in industrially raised layers [92]. Different production systems use antibiotics differently and this accounted for the varied resistance profiles. In 2012, a study found higher resistance to ciprofloxacin in cattle than broilers in the Venda Region in the north of South Africa [91]. In chicken, *C. jejuni* showed a resistance of 29% to ciprofloxacin while in cattle the resistance was 33.3%. However, resistance to erythromycin was higher at 56.7% and 42.9% in chicken and cattle respectively. Resistance to ciprofloxacin among chicken was due to the use of sarafloxacin and enrofloxacin on poultry farms. Additionally, Bester and Essack [28] found varying resistances to common antibiotics used in broilers and layers in Kwazulu Natal. For instance, resistance to tetracycline was 98.2% and 100% for broilers and layers respectively. Resistance to gentamicin was equally high at 98% for broilers and 81% for layers. Thirdly, multi-resistance was found in 23% and 43% of the isolates from broilers and layers, respectively.

4.4. North Africa

In Algeria, isolated strains from turkeys showed resistance to tetracycline at 81.3%, ciprofloxacin at 75.0%, and erythromycin at 25.0%. 96.9% of the isolates were multidrug resistant and 18 drug resistance patterns were identified [77]. The authors in the previous study argued that the prudent use of fluoroquinolones contributed to the high resistance rates recorded in ciprofloxacin. The continuous use of fluoroquinolones in poultry production selects for fluoroquinolone-resistant mutants which leads to the emergence of resistant *Campylobacter*. The selection pressure presented by different antibiotics accounted for the varying rate of resistance and multidrug resistance trends recorded. Meryem, Zehor, Fares, Sadjia and Amina [29], also discovered that the isolates from livestock showed resistance of 91.6% to ciprofloxacin, 88.54% to erythromycin, and 44.7% to tetracycline in the middle area of Algeria. Resistance to ciprofloxacin and erythromycin were 91.6% and 88.54% respectively while that to tetracycline was 44.7%.

Reports from Egypt indicated that *Campylobacter* isolates from poultry showed a 58.82% resistance to erythromycin and tetracycline; whereas isolates from humans showed a resistance of 75% to tetracycline and 62.5% to erythromycin [57]. The inexpensive cost and broad-spectrum properties of tetracycline resulted in the widespread use in both humans and animals and consequently, the selective pressure led to the emergence of resistant genes. In addition, *C. jejuni* and *C. coli* isolates from diarrhoeal patients in Cairo showed 40% and 24% resistance to nalidixic acid, 6% and 24% to tetracycline and 9% and 10% resistance to erythromycin respectively [59].

5. Management Strategies: Opportunities and Challenges

A holistic, multiple intervention One Health approach is required to better understand, prevent, and control *Campylobacter* and its related infections. One Health recognizes that the health of people is connected to the health of animals and the environment. It is a collaborative, multisectoral, and transdisciplinary approach—working at the local, regional, national, and global levels—with the goal of achieving optimal health outcomes recognising the interconnection between people, animals, plants, and their shared environment (readers unfamiliar with the One Health concept could consult the following CDC website for an introduction: https://www.cdc.gov/onehealth/index.html). In the African context, management of *Campylobacter* can be dealt with at the domestic, farm, processing, and policy levels.

At the domestic setting level, several studies across Africa indicated a significant association between *Campylobacter* enteritis and contact with both animals [32,41,42,93,94] and diarrhoea sick persons [32]. This creates a public health concern since agriculture, particularly the livestock sector, is a major source of livelihood in Africa. In rural areas where sometimes access to adequate medical intervention is unavailable the situation becomes more critical. Furthermore, the communal living culture in Africa tends to have healthy family members taking care of sick ones, and those with diarrhoea are no exceptions from receiving family care. It is important to minimise human-animal contact, practice personal and environmental hygiene, and seek proper medical care for sick persons in a household in order to minimise risks of transmission.

Animal production and management systems play an important part in *Campylobacter* control and must be carefully considered. Studies in Europe estimated that a three log unit reduction of *Campylobacter* load in the ceca of poultry would result to a more than 90% reduction of human infections attributed to poultry meat consumption [22]. Additionally, the animal production system influences antimicrobial usage and this has implications for antibiotic resistance. According to Jonker and Picard [84], in intensive poultry and pig rearing systems the use of oral antibiotics is essential to maintain health hence there is a high risk for the *Campylobacter* in the intestinal tract of food animals to develop resistance to commonly used antibiotics. In poultry flocks, the risk factors for *Campylobacter* introduction include the partial depopulation of flocks at several occasions, the increased age of sale, inadequate washing and disinfection of poultry houses, and poor levels of biosecurity measures such as absence of a footbath at the entrance and poor rodent control [95]. Treatment of drinking water and litter, rodent control on farms, and effective biosecurity are important management interventions to be employed on the farm [96]. Applying biosecurity interventions at poultry production sites has resulted in different levels of success in different countries [6,14,97]. Such variation may be attributed to differences in the *Campylobacter* loads in the poultry chain and environment. Hence, the effectiveness of biosecurity-related interventions in primary production should be based on a good understanding of the regional risk factors at the farm level.

In addition to primary interventions at the farm level, there is a need to apply interventions at the slaughter and processing levels in order to reduce the contamination of poultry meat meant for human consumption. Freezing of contaminated poultry carcasses is a reliable intervention to achieve a 1 to 2-log reduction of *Campylobacter* counts. The compulsory freezing of processed broilers from *Campylobacter*-positive broiler flocks in Iceland resulted in a substantially reduced number of human cases of *Campylobacter* enteritis and is currently being used on a voluntary basis in Norway, Sweden, and Denmark [14,22]. However, in Africa, freezing might not be a feasible option in many countries due to the added cost of energy used for freezing operations as well as challenges in keeping a sustainable cold chain in several countries with limited infrastructure. Many consumers across the African continent prefer to buy fresh poultry meat with no change in product quality [24].

Chemical decontamination can also be an effective intervention for reducing *Campylobacter* load on food animal carcasses, and the feasibility of this option could be appealing in the African context. Chlorine, chlorine dioxide, acidified sodium chlorite, trisodium phosphate, and peroxyacid are typically used in poultry processing in the United States and Australia either as sprays or washes for online reprocessing or added to the chill water tank [7,9]. However, pathogens decontamination using chemicals and its application to poultry carcasses should be well regulated and monitored, in order to avoid excessive use. Also the use of chemical decontamination should not be considered as a replacement, rather complimentary, to good processing practices in poultry abattoirs [14].

The lack of national surveillance data hinders the adequate assessment of the public health impact and burden of disease [7,22]. Most African countries have no national surveillance programs on *Campylobacter*. Therefore, getting the accurate burden of the disease is difficult [24]. Adopting multisector collaborations would help strengthen the disease surveillance system, enhance laboratory capacity, and support the implementation of prevention and control strategies; it would further enhance public health and veterinary laboratories and create intersectorial linkages to tackle zoonotic diseases [97]. In spite of technological advances, laboratories in many African countries still face challenges with isolation and identification of *Campylobacter* in food and clinical samples. Inadequately trained personnel, poor laboratory infrastructure, low funding for research on foodborne pathogens like *Campylobacter* are also major limiting factors in many countries across Africa.

6. Conclusions

Zoonotic pathogens such as *Campylobacter* cause disease and death, resulting in enormous social and economic losses. Significant gaps exist in the epidemiological understanding of *Campylobacter* in Africa, and its role as a diarrhoeal agent needs more attention. The present review highlights the variability in the reported occurrence of *Campylobacter* in humans and animal food sources across different countries and regions in Africa. *Campylobacter* infection is particularly prevalent in the paediatric population and has been isolated from farm animals, particularly poultry, and foods of animal origin. The reported prevalence of *Campylobacter* in children under the age of five years ranges from 2% in Sudan to 21% in South Africa. To better understand the magnitude of the campylobacteriosis burden, future research is required to evaluate the under-reporting of diarrhoea incidents at national and continental levels; this cannot be achieved without enhancing local capacities for disease surveillance and monitoring.

A holistic, multiple intervention One Health approach is required to better understand, prevent, and control *Campylobacter* in the African context. The management of the foodborne transmission of *Campylobacter* can be dealt with at the domestic, farm, processing, and policy levels. In poultry, the present review points that the prevalence ranges from 14.4% in Ghana to 96% in Algeria. This review also highlights the alarming trend across several countries in Africa of increased *Campylobacter* resistance to clinically important antimicrobials, such as ciprofloxacin and erythromycin, in humans and food animals.

In our opinion, and based on the evidence gathered in this review, we believe that *Campylobacter* infection is predicted to emerge further as a serious challenge that African nations will face in the near future. Children in low-resource settings countries across Africa will be suffering the most, given the high burden of infection, combined with the growing trend in *Campylobacter* resistance to clinically relevant antimicrobial. We still know little about from what, how, and where children contract infection. What role do domestic animals, which are known reservoirs of *Campylobacter*, play in transmission? Does infection result from fecal contamination in the environment and how long does *Campylobacter* survive in the environment? Although exposure to poultry may be important, identified determinants are varied across different regions in Africa. Given the paucity of current data, further research on *Campylobacter* in Africa is warranted. Epidemiological studies on risk factors and exposure routes would assist in devising appropriate interventions and strategies. However, it is important to prioritise the reduction of risk factors and exposure routes in all settings as a first step in management.

Author Contributions: Conceptualisation, I.H.; Methodology, I.H. and N.A.-B.; Writing—original draft, N.A.-B., A.P.-H.K., S.A. and I.H.; Writing—review & editing, N.A.-B., A.P.-H.K. and I.H.

Funding: This research received no external funding. Nikki Asuming-Bediako is a PhD student supported by a scholarship from Murdoch University.

Acknowledgments: This research is supported by an Australian Government Research Training Program (RTP) Scholarship awarded to Nikki Asuming-Bediako in support of her PhD study at Murdoch University.

Conflicts of Interest: The authors declare no conflict of interest.

References

1. Gharst, G.; Oyarzabal, O.A.; Hussain, S.K. Review of current methodologies to isolate and identify *Campylobacter* spp. from foods. *J. Microbiol. Methods* **2013**, *95*, 84–92. [CrossRef] [PubMed]

2. Adams, M.R.; Moss, M.O. *Food Microbiology*; The Royal Society of Chemistry: Cambridge, UK, 2008.

3. Padungton, P.; Kaneene, J.B. *Campylobacter spp.* in Humans, Chickens, Pigs and Their Antimicrobial Resistance. *J. Vet. Med. Sci* **2003**, *65*, 161–170. [CrossRef] [PubMed]

4. Elbrissi, A.; Sabeil, Y.A.; Khalifa, K.A.; Enan, K.; Khair, O.M.; El Hussein, A.M. Isolation, identification and differentiation of *Campylobacter* spp. using multiplex PCR assay from goats in Khartoum State, Sudan. *Trop. Anim. Health Prod.* **2017**, *49*, 575–581. [CrossRef] [PubMed]

5. Tafa, B.; Sewunet, T.; Tassew, H.; Asrat, D. Isolation and Antimicrobial Susceptibility Patterns of Campylobacter Species among Diarrheic Children at Jimma, Ethiopia. *Int. J. Bacteriol.* **2014**, *2014*, 560617. [CrossRef] [PubMed]

6. Goualie, G.B.; Essoh, E.A.; Elise Solange, K.N.; Natalie, G.; Souleymane, B.; Lamine Sebastien, N.; Mireille, D. Prevalence and Antimicrobial Resistance of Thermophilic Campylobacter Isolated from Chicken in Cote d'Ivoire. *Int. J. Microbiol.* **2012**, *2012*, 150612.

7. Butzler, J.P. Campylobacter, from obscurity to celebrity. *Clin. Microbiol. Infect.* **2004**, *10*, 868–876. [CrossRef] [PubMed]

8. Whiley, H.; van den Akker, B.; Giglio, S.; Bentham, R. The role of environmental reservoirs in human campylobacteriosis. *Int. J. Environ. Res. Public Health* **2013**, *10*, 5886–5907. [CrossRef] [PubMed]

9. WHO. *WHO Estimates of the Global Burden of Foodborne Diseases*; WHO: Geneva, Switzerland, 2015; p. 268.

10. Komba, E.V.G. Human and Animal Thermophilic Campylobacter infections in East African countries: Epidemiology and Antibiogram. *Biomed. J. Sci. Tech. Res.* **2017**, *16*, 17. [CrossRef]

11. Kaakoush, N.O.; Castano-Rodriguez, N.; Mitchell, H.M.; Man, S.M. Global Epidemiology of Campylobacter Infection. *Clin. Microbiol. Rev.* **2015**, *28*, 687–720. [CrossRef]

12. Havelaar, A.H.; van Pelt, W.; Ang, C.W.; Wagenaar, J.A.; van Putten, J.P.; Gross, U.; Newell, D.G. Immunity to Campylobacter: Its role in risk assessment and epidemiology. *Crit. Rev. Microbiol.* **2009**, *35*, 1–22. [CrossRef]

13. Ahs, J.W.; Tao, W.; Löfgren, J.; Forsberg, B.C. Diarrheal Diseases in Low- and Middle-Income Countries: Incidence, Prevention and Management. *Open Infect. Dis. J.* **2010**, *4*, 113–124. [CrossRef]

14. Humphrey, T.; O'Brien, S.; Madsen, M. Campylobacters as zoonotic pathogens: A food production perspective. *Int. J. Food Microbiol.* **2007**, *117*, 237–257. [CrossRef] [PubMed]

15. Dasti, J.I.; Tareen, A.M.; Lugert, R.; Zautner, A.E.; Gross, U. Campylobacter jejuni: A brief overview on pathogenicity-associated factors and disease-mediating mechanisms. *Int. J. Med. Microbiol.* **2010**, *300*, 205–211. [CrossRef] [PubMed]

16. Park, S.F. The physiology of Campylobacter species and its relevance to their role as foodborne pathogens. *Int. J. Food Microbiol.* **2000**, *74*, 177–188. [CrossRef]

17. Konkel, M.E.; Garvis, S.G.; Tipton, S.L.; Anderson, D.E.J.; Cieplak, W.J. Identification and molecular cloning of a gene encoding a fibronectin-binding protein (CadF) from *Campylobacter jejuni*. *Mol. Microbiol.* **1997**, *24*, 953–963. [CrossRef] [PubMed]

18. Carvalho, A.C.; Ruiz-Palacios, G.M.; Ramos-Cervantes, P.; Cervantes, L.E.; Jiang, X.; Pickering, L.K. Molecular characterization of invasive and noninvasive Campylobacter jejuni and Campylobacter coli isolates. *J. Clin. Microbiol.* **2001**, *39*, 1353–1359. [CrossRef] [PubMed]

19. Hickey, T.E.; Mcveigh, A.L.; Scott, D.A.; Michielutti, R.E.; Bixby, A.; Carroll, S.A.; Bourgeois, L.A.; Guerry, P. *Campylobacter jejuni* Cytolethal Distending Toxin Mediates Release of Interleukin-8 from Intestinal Epithelial Cells. *Infect. Immun.* **2000**, *68*, 6535–6541. [CrossRef]

20. Tresse, O.; Alvarez-Ordonez, A.; Connerton, I.F. Editorial: About the Foodborne Pathogen Campylobacter. *Front. Microbiol.* **2017**, *8*, 1908. [CrossRef]

21. Pesci, E.C.; Cottle, D.L.; Pickett, C.L. Genetic, Enzymatic, and Pathogenic Studies of the Iron Superoxide Dismutase of Campylobacter jejuni. *Infect. Immun.* **1994**, *62*, 2687–2694.

22. WHO. *The Global View of Campylobacteriosis*; World Health Organization: Geneva, Switzerland, 2013; p. 69.
23. Samuel, S.O.; Aboderin, A.O.; Akanbi, A.A., II; Adegboro, B.; Smith, S.I.; Coker, A.O. *Campylobacter* enetritis in Ilorin, Nigeria. *East Afr. Med. J.* **2006**, *83*, 478–484.
24. Coker, A.O.; Isokpehi, R.D.; Thomas, B.N.; Amisu, K.O.; Obi, C.L. Human Campylobacteriosis in Developing Countries. *Emerg. Infect. Dis.* **2002**, *8*, 237–243. [CrossRef] [PubMed]
25. Adekunle, O.C.; Coker, A.O.; Kolawole, D.O. Antibiotic susceptibility pattern of strains of *Campylobacter coli* isolated in Osogbo, Nigeria. *Biol. Med.* **2009**, *1*, 20–23.
26. Moyane, J.N.; Jideani, A.I.O.; Aiyegoro, O.A. Antibiotics usage in food-producing animals in South Africa and impact on human: Antibiotic resistance. *Afr. J. Microbiol. Res.* **2013**, *7*, 2990–2997.
27. Mshana, S.E.; Joloba, M.; Kakooza, A.; Kaddu-Mulindwa, D. *Campylobacter* spp among Children with acute diarrhea attending Mulago hospital in Kampala—Uganda. *Afr. Health Sci.* **2009**, *9*, 201–205. [PubMed]
28. Bester, L.A.; Essack, S.Y. Prevalence of antibiotic resistance in Campylobacter isolates from commercial poultry suppliers in KwaZulu-Natal, South Africa. *J. Antimicrob. Chemother.* **2008**, *62*, 1298–1300. [CrossRef] [PubMed]
29. Meryem, G.; Zehor, G.; Fares, A.; Sadjia, M.; Amina, H. Campylobacter in sheep, calves and broiler chickens in the central region of Algeria: Phenotypic and antimicrobial resistance profiles. *Afr. J. Microbiol. Res.* **2016**, *10*, 1662–1667. [CrossRef]
30. UNSD Methodology. Available online: www.unstats.un.org/unsd/methodology/m49/ (accessed on 8 April 2019).
31. Lengerh, A.; Moges, F.; Unakal, C.; Anagaw, B. Prevalence, associated risk factors and antimicrobial susceptibility pattern of Campylobacter species among under five diarrheic children at Gondar University Hospital, Northwest Ethiopia. *BMC Pediatr.* **2013**, *13*, 82. [CrossRef] [PubMed]
32. Mitike, G.; Kassu, A.; Genetu, A.; Nigussie, D. *Campylobacter* enteritis among children in Dembia District, Northwest Ethiopia. *East Afr. Med. J.* **2000**, *77*, 654–657. [CrossRef]
33. Swierczewski, B.E.; Odundo, E.A.; Koech, M.C.; Ndonye, J.N.; Kirera, R.K.; Odhiambo, C.P.; Cheruiyot, E.K.; Shaffer, D.N.; Ombogo, A.N.; Oak, E.V. Enteric pathogen surveillance in a case-control study of acute diarrhea in Kisii Town, Kenya. *J. Med. Microbiol.* **2013**, *62 Pt 11*, 1774–1776. [CrossRef]
34. Randremanana, R.; Randrianirina, F.; Gousseff, M.; Dubois, N.; Razafindratsimandresy, R.; Hariniana, E.R.; Garin, B.; Randriamanantena, A.; Rakotonirina, H.C.; Ramparany, L.; et al. Case-control study of the etiology of infant diarrheal disease in 14 districts in Madagascar. *PLoS ONE* **2012**, *7*, e44533. [CrossRef] [PubMed]
35. Mason, J.; Iturriza-Gomara, M.; O'Brien, S.J.; Ngwira, B.M.; Dove, W.; Maiden, M.C.; Cunliffe, N.A. Campylobacter infection in children in Malawi is common and is frequently associated with enteric virus co-infections. *PLoS ONE* **2013**, *8*, e59663. [CrossRef] [PubMed]
36. Mandomando, M.I.; Macete, V.E.; Ruiz, J.; Sanz, S.; Abacassamo, F.; Valles, X.; Sacarlal, J.; Navia, M.M.; Vila, J.; Pedro, L.P.; et al. Etiology of Diarrhea in Children Younger than 5 years Of Age Admitted in A Rural Hospital of Southern Mozambique. *Am. J. Trop. Med. Hyg.* **2007**, *76*, 522–527. [CrossRef] [PubMed]
37. Komba, E.V.; Mdegela, R.H.; Msoffe, P.L.; Nielsen, L.N.; Ingmer, H. Prevalence, Antimicrobial Resistance and Risk Factors for Thermophilic Campylobacter Infections in Symptomatic and Asymptomatic Humans in Tanzania. *Zoonoses Public Health* **2015**, *62*, 557–568. [CrossRef] [PubMed]
38. Deogratias, A.-P.; Mushi, M.F.; Paterno, L.; Tappe, D.; Seni, J.; Kabymera, R.; Kidenya, B.R.; Mshana, S.E. Prevalence and determinants of *Campylobacter* infection among under five children with acute watery diarrhea in Mwanza, North Tanzania. *Arch. Public Health* **2014**, *72*, 17. [CrossRef] [PubMed]
39. Mdegela, R.H.; Nonga, H.E.; Ngowi, H.A.; Kazwala, R.R. Prevalence of Thermophilic *Campylobacter* Infections in Humans, Chickens and Crows in Morogoro, Tanzania. *J. Vet. Med.* **2006**, *53*, 116–121. [CrossRef] [PubMed]
40. Pelkonen, T.; Dos Santos, M.D.; Roine, I.; Dos Anjos, E.; Freitas, C.; Peltola, H.; Laakso, S.; Kirveskari, J. Potential Diarrheal Pathogens Common Also in Healthy Children in Angola. *Pediatr. Infect. Dis. J.* **2018**, *37*, 424–428. [CrossRef] [PubMed]
41. Sangaré, L.; Nikiéma, A.K.; Zimmermann, S.; Sanou, I.; Congo-Ouédraogo, M.; Diabaté, A.; Diandé, S.; Guissou, P.I. *Campylobacter Spp.* Epidemiology and Antimicrobial Susceptibility in a Developing Country, Burkina Faso (West Africa). *Afr. J. Clin. Exp. Microbiol.* **2012**, *13*, 110–117. [CrossRef]
42. Karikari, A.B.; Obiri-Danso, K.; Frimpong, E.H.; Krogfelt, K.A. Antibiotic Resistance in Campylobacter Isolated from Patients with Gastroenteritis in a Teaching Hospital in Ghana. *Open J. Med. Microbiol.* **2017**, *7*, 1–11. [CrossRef]

43. Molbak, K.; Hojlyng, N.; Gaarslev, K. High prevalence of campylobacter excretors among Liberian children related to environmental conditions. *Epidem. Inf.* **1988**, *100*, 227–237. [CrossRef]

44. Nwankwo, I.O.; Faleke, O.O.; Salihu, M.D.; Magaji, A.A.; Musa, U.; Garba, J. Epidemiology of Campylobacter species in poultry and humans in the four agricultural zones of Sokoto State, Nigeria. *J. Public Health Epidemiol.* **2016**, *8*, 184–190.

45. Nwankwo, I.O.; Faleke, O.O.; Salihu, M.D.; Magaji, A.A.; Shu-Yun, Z. Prevalence of Campylobacter Species in Out-patients And Pregnant Women Attending Government Clinics in Sokoto State, Nigeria. *J. Prev. Med. Care* **2016**, *1*, 8–15.

46. Ohanu, M.E.; Offune, J. The prevalence of Campylobacter in childhood diarrhea in Enugu State of Nigeria. *J. Commun. Dis.* **2009**, *41*, 117–120. [PubMed]

47. Samie, A.; Obi, C.L.; Barrett, L.J.; Powell, S.M.; Guerrant, R.L. Prevalence of Campylobacter species, Helicobacter pylori and Arcobacter species in stool samples from the Venda region, Limpopo, South Africa: Studies using molecular diagnostic methods. *J. Infect.* **2007**, *54*, 558–566. [CrossRef] [PubMed]

48. Samie, A.; Ramalivhana, J.; Igumbor, E.O.; Obi, C.L. Prevalence, haemolytic and haemagglutination activities and antibiotic susceptibility profiles of *Campylobacter* spp. isolated from human diarrhoeal stools in Vhembe District, South Africa. *J. Health Popul. Nutr.* **2007**, *25*, 406–413. [PubMed]

49. Lastovica, A.J. *Emerging Campylobacter* spp.: The Tip of the Iceberg. *Clin. Microbiol. Newsl.* **2006**, *28*, 49–56. [CrossRef]

50. Obi, C.L.; Bessong, P.O. Diarrhoeagenic bacterial pathogens in HIV-positive patients with diarrhoea in rural communities of Limpopo Province, South Africa. *J. Health Popul. Nutr.* **2002**, *20*, 230–234.

51. Mackenjee, M.K.; Coovadia, Y.M.; Coovadia, H.M.; Hewitt, J.; Robins-Browne, R.M. Aetiology of diarrhoea in adequately nourished young African children in Durban, South Africa. *Ann. Trop. Paediatr.* **1984**, *4*, 183–187. [CrossRef]

52. Abushahba, M.F.N. Prevalence of Zoonotic Species of Campylobacter in Broiler Chicken and Humans in Assiut Governorate, Egypt. *Approaches Poult. Dairy Vet. Sci.* **2018**, *3*, 1–9. [CrossRef]

53. Sainato, R.; ElGendy, A.; Poly, F.; Kuroiwa, J.; Guerry, P.; Riddle, M.S.; Porter, C.K. Epidemiology of Campylobacter Infections among Children in Egypt. *Am. J. Trop. Med. Hyg.* **2018**, *98*, 581–585. [CrossRef]

54. El-Tras, W.F.; Holt, H.R.; Tayel, A.A.; El-Kady, N.N. Campylobacter infections in children exposed to infected backyard poultry in Egypt. *Epidemiol. Infect.* **2015**, *143*, 308–315. [CrossRef]

55. Omara, S.T.; Fadaly, H.A.E.; Barakat, A.M.A. Public Health Hazard of Zoonotic Campylobacter jejuni Reference to Egyptian Regional and Seasonal Variations. *Res. J. Microbiol.* **2015**, *10*, 343–354. [CrossRef]

56. Awadallah, M.A.I.; Ahmed, H.A.; El-Gedawy, A.A.; Saad, A.M. Molecular identification of C. jejuni and C. coli in chicken and humans, at Zagazig, Egypt, with reference to the survival of C. jejuni in chicken meat at refrigeration and freezing temperatures. *Int. Food Res. J.* **2014**, *21*, 1801–1812.

57. Hassanain, N.A. Antimicrobial Resistant *Campylobacter jejuni* Isolated from Humans and Animals in Egypt. *Glob. Vet.* **2011**, *6*, 195–200.

58. Rao, M.R.; Naficy, A.B.; Savarino, S.J.; Abu-Elyazeed, R.; Wierzba, T.; Peruski, L.; Abdel-Messih, I.A.; Frenck, R.R.; Clemens, J.D. Pathogenicity and Convalescent Excretion of Campylobacter in Rural Egyptian Children. *Am. J. Epidemiol.* **2001**, *154*, 166–173. [CrossRef] [PubMed]

59. Wasfy, M.O.; Oyofo, B.A.; David, J.C.; Ismail, T.F.; El-Gendy, A.M.; Mohran, Z.S.; Sultan, Y.; Peruski, L.F. Isolation and Antibiotic Susceptibility of Salmonella, Shigella, and Campylobacter from Acute Enteric Infections in Egypt. *J. Health Popul. Nutr.* **2000**, *18*, 33–38. [PubMed]

60. Saeed, A.; Abd, H.; Sandstrom, G. Microbial aetiology of acute diarrhoea in children under five years of age in Khartoum, Sudan. *J. Med. Microbiol.* **2015**, *64 Pt 4*, 432–437. [CrossRef]

61. Heikema, A.P.; Islam, Z.; Horst-Kreft, D.; Huizinga, R.; Jacobs, B.C.; Wagenaar, J.A.; Poly, F.; Guerry, P.; van Belkum, A.; Parker, C.T.; et al. Campylobacter jejuni capsular genotypes are related to Guillain-Barre syndrome. *Clin. Microbiol. Infect.* **2015**, *21*, 852.e1–852.e9. [CrossRef]

62. Paudyal, N.; Anihouvi, V.; Hounhouigan, J.; Matsheka, M.I.; Sekwati-Monang, B.; Amoa-Awua, W.; Atter, A.; Ackah, N.B.; Mbugua, S.; Asagbra, A.; et al. Prevalence of foodborne pathogens in food from selected African countries—A meta-analysis. *Int. J. Food Microbiol.* **2017**, *249*, 35–43. [CrossRef]

63. Woldemariam, T.; Asrat, D.; Zewde, G. Prevalence of Thermophilic Campylobacter species in carcasses from sheep and goats in an abattoir in Debre Zeit area, Ethiopia. *Ethiop. J. Health Dev.* **2009**, *23*, 229–232. [CrossRef]

64. Osano, O.; Arimi, S.M. Retail Poultry and Beef as Sources of *Campylobacter jejuni*. *East Afr. Med. J.* **1999**, *76*, 141–143.

65. Kashoma, I.P.; Kassem, I.I.; John, J.; Kessy, B.M.; Gebreyes, W.; Kazwala, R.R.; Rajashekara, G. Prevalence and Antimicrobial Resistance of Campylobacter Isolated from Dressed Beef Carcasses and Raw Milk in Tanzania. *Microb. Drug Resist.* **2016**, *22*, 40–52. [CrossRef] [PubMed]

66. Nonga, H.E.; Muhairwa, A.P. Prevalence and antibiotic susceptibility of thermophilic Campylobacter isolates from free range domestic duck (Cairina moschata) in Morogoro municipality, Tanzania. *Trop. Anim. Health Prod.* **2010**, *42*, 165–172. [CrossRef] [PubMed]

67. Nonga, H.E.; Sells, P.; Karimuribo, E.D. Occurrences of thermophilic Campylobacter in cattle slaughtered at Morogoro municipal abattoir, Tanzania. *Trop. Anim. Health Prod.* **2010**, *42*, 73–78. [CrossRef]

68. Nzouankeu, A.; Ngandjio, A.; Ejenguele, G.; Njine, T.; Wouafo, M.N. Multiple contaminations of chickens with Campylobacter, Escherichia coli and Salmonella in Yaounde (Cameroon). *J. Infect. Dev. Ctries.* **2010**, *4*, 583–586. [PubMed]

69. Mpalang, R.K.; Boreux, R.; Melin, P.; Akir Ni Bitiang, K.; Daube, G.; De Mol, P. Prevalence of Campylobacter among goats and retail goat meat in Congo. *J. Infect. Dev. Ctries.* **2014**, *8*, 168–175. [CrossRef] [PubMed]

70. Kagambèga, A.; Thibodeau, A.; Trinetta, V.; Soro, D.K.; Sama, F.N.; Bako, É.; Bouda, C.S.; Wereme N'Diaye, A.; Fravalo, P.; Barro, N. *Salmonella* spp. and *Campylobacter* spp. in poultry feces and carcasses in Ouagadougou, Burkina Faso. *Food Sci. Nutr.* **2018**, *6*, 1601–1606. [CrossRef] [PubMed]

71. Karikari, A.B.; Obiri-Danso, K.; Frimpong, E.H.; Krogfelt, K.A. Multidrug resistant *Campylobacter* in faecal and carcasses of commercially produced poultry. *Afr. J. Microbiol. Res.* **2017**, *11*, 271–277.

72. Karikari, A.B.; Obiri-Danso, K.; Frimpong, E.H.; Krogfelt, K.A. Antibiotic Resistance of Campylobacter Recovered from Faeces and Carcasses of Healthy Livestock. *BioMed Res. Int.* **2017**, *2017*, 4091856. [CrossRef]

73. Salihu, M.D.; Junaidu, A.U.; Magaji, A.A.; Rabiu, Z.M. Study of *Campylobacter* in Raw Cow Milk in Sokoto State, Nigeria. *Br. J. Dairy Sci.* **2010**, *1*, 1–5.

74. Salihu, M.D.; Junaidu, A.U.; Magaji, A.A.; Abubakar, M.B.; Adamu, A.Y.; Yakubu, A.S. Prevalence of Campylobacter in poultry meat in Sokoto, Northwestern Nigeria. *J. Public Health Epidemiol.* **2009**, *1*, 041–045.

75. Cardinale, E.; Perrier Gros-Claude, J.D.; Tall, F.; Cissé, M.; Guèye, E.F.; Salvat, G. Prevalence of Salmonella and Campylobacter in Retail Chicken Carcasses in Senegal. *Rev. d'élevage Méd. Vét. Pays Trop.* **2003**, *56*, 13–16.

76. van Nierop, W.; Duse, A.G.; Marais, E.; Aithma, N.; Thothobolo, N.; Kassel, M.; Stewart, R.; Potgieter, A.; Fernandes, B.; Galpin, J.S.; et al. Contamination of chicken carcasses in Gauteng, South Africa, by Salmonella, Listeria monocytogenes and Campylobacter. *Int. J. Food Microbiol.* **2005**, *99*, 1–6. [CrossRef]

77. Bouhamed, R.; Bouayad, L.; Messad, S.; Zenia, S.; Naim, M.; Hamdi, T.-M. Sources of contamination, prevalence, and antimicrobial resistance of thermophilic Campylobacter isolated from turkeys. *Vet. World* **2018**, *11*, 1074–1081. [CrossRef] [PubMed]

78. Laidouci, A.A.H.; Mouffok, F.; Hellal, A. Detection of Campylobacter in poultry in Algeria: Antibacterial profile determination. *Rev. Méd. Vét.* **2013**, *164*, 307–311.

79. El-Zamkan, M.A.; Hameed, K.G. Prevalence of Campylobacter jejuni and Campylobacter coli in raw milk and some dairy products. *Vet. World* **2016**, *9*, 1147–1151. [CrossRef] [PubMed]

80. El-Sharoud, W.M. Prevalence and survival of Campylobacter in Egyptian dairy products. *Food Res. Int.* **2009**, *42*, 622–626. [CrossRef]

81. Jouahri, M.; Asehraou, A.; Karib, H.; Hakkou, A.; Touhami, M. Prevalence and Control of Thermotolerant Campylobacter Species in Raw Poultry Meat in Morocco. *MESO* **2007**, *9*, 262–267.

82. Moore, J.E.; Barton, M.D.; Blair, I.S.; Corcoran, D.; Dooley, J.S.; Fanning, S.; Kempf, I.; Lastovica, A.J.; Lowery, C.J.; Matsuda, M.; et al. The epidemiology of antibiotic resistance in Campylobacter. *Microbes Infect.* **2006**, *8*, 1955–1966. [CrossRef]

83. de Vries, S.P.W.; Vurayai, M.; Holmes, M.; Gupta, S.; Bateman, M.; Goldfarb, D.; Maskell, D.J.; Matsheka, M.I.; Grant, A.J. Phylogenetic analyses and antimicrobial resistance profiles of *Campylobacter* spp. from diarrhoeal patients and chickens in Botswana. *PLoS ONE* **2018**, *13*, e0194481. [CrossRef]

84. Jonker, A.; Picard, J.A. Antimicrobial susceptibility in thermophilic Campylobacter species isolated from pigs and chickens in South Africa. *J. S. Afr. Vet. Assoc.* **2010**, *81*, 228–236. [CrossRef]

85. Ghunaim, H.; Behnke, J.M.; Aigha, I.; Sharma, A.; Doiphode, S.H. Analysis of resistance to antimicrobials and presence of virulence/stress response genes in Campylobacter isolates from patients with severe diarrhoea. *PLoS ONE* **2015**, *10*, e0119268. [CrossRef] [PubMed]

86. Iovine, N.M. Resistance mechanisms in Campylobacter jejuni. *Virulence* **2013**, *4*, 230–240. [CrossRef] [PubMed]

87. Nguyen, T.N.; Hotzel, H.; Njeru, J.; Mwituria, J.; El-Adawy, H.; Tomaso, H.; Neubauer, H.; Hafez, H.M. Antimicrobial resistance of Campylobacter isolates from small scale and backyard chicken in Kenya. *Gut Pathog.* **2016**, *8*, 39. [CrossRef] [PubMed]

88. Salihu, M.D.; Junaidu, A.U.; Magaji, A.A.; Yakubu, Y. Prevalence and Antimicrobial Resistance of Thermophilic Campylobacter Isolates from Commercial Broiler Flocks in Sokoto, Nigeria. *Res. J. Vet. Sci.* **2012**, *5*, 51–58. [CrossRef]

89. Reddy, S.; Zishiri, O.T. Detection and prevalence of antimicrobial resistance genes in *Campylobacter* spp. isolated from chickens and humans. *Onderstepoort J. Vet. Res.* **2017**, *84*, e1–e6. [CrossRef] [PubMed]

90. Shobo, C.O.; Bester, L.A.; Baijnath, S.; Somboro, A.M.; Peer, A.K.; Essack, S.Y. Antibiotic resistance profiles of Campylobacter species in the South Africa private health care sector. *J. Infect. Dev. Ctries.* **2016**, *10*, 1214–1221. [CrossRef]

91. Uaboi-Egbenni, P.O.; Bessong, P.O.; Samie, A.; Obi, C.L. Potentially pathogenic Campylobacter species among farm animals in rural areas of Limpopo province, South Africa: A case study of chickens and cattles. *Afr. J. Microbiol. Res.* **2012**, *6*, 2835–2843. [CrossRef]

92. Bester, L.A.; Essack, S.Y. Observational study of the prevalence and antibiotic resistance of *Campylobacter* spp. from different poultry production systems in KwaZulu-Natal, South Africa. *J. Food Prot.* **2012**, *75*, 154–159. [CrossRef]

93. Wierzba, T.F.; Abdel-Messih, I.A.; Gharib, B.; Baqar, S.; Hendaui, A.; Khalil, I.; Omar, T.A.; Khayat, H.E.; Putnam, S.D.; Sanders, J.W.; et al. Campylobacter infection as a trigger for Guillain-Barre syndrome in Egypt. *PLoS ONE* **2008**, *3*, e3674. [CrossRef]

94. Klous, G.; Huss, A.; Heederik, D.J.J.; Coutinho, R.A. Human-livestock contacts and their relationship to transmission of zoonotic pathogens, a systematic review of literature. *One Health* **2016**, *2*, 65–76. [CrossRef]

95. Mageto, L.M.; Ombui, J.N.; Mutua, F.K. Prevalence and risk factors for Campylobacter infection of chicken in peri-urban areas of Nairobi, Kenya. *J. Dairy Vet. Anim. Res.* **2018**, *7*, 22–27. [CrossRef]

96. Orhan, S.; Issmat, I.K.; Zhangqi, S.; Jun, L.; Gireesh, R.; Qijing, Z. Campylobacter in Poultry: Ecology and Potential Interventions. *Avian Dis.* **2015**, *59*, 185–200.

97. Pieracci, E.G.; Hall, A.J.; Gharpure, R.; Haile, A.; Walelign, E.; Deressa, A.; Bahiru, G.; Kibebe, M.; Walke, H.; Belay, E. Prioritizing zoonotic diseases in Ethiopia using a one health approach. *One Health* **2016**, *2*, 131–135. [CrossRef] [PubMed]

 *pathogens*

Review

Childhood Diarrhoea in the Eastern Mediterranean Region with Special Emphasis on Non-Typhoidal Salmonella at the Human–Food Interface

Ali Harb [1,2], **Mark O'Dea** [1], **Sam Abraham** [1] and **Ihab Habib** [1,3,*]

1   College of Science, Health, Education and Engineering, Murdoch University, Perth 6150, Australia; aliharb8@yahoo.com (A.H.); m.o'dea@murdoch.edu.au (M.O.); s.abraham@murdoch.edu.au (S.A.)
2   Thi-Qar Public Health Division, Ministry of Health, Thi-Qar 64007, Iraq
3   High Institute of Public Health, Alexandria University, Alexandria 21516, Egypt
*   Correspondence: i.habib@murdoch.edu.au; Tel.: +61-89-360-2434

Received: 12 April 2019; Accepted: 1 May 2019; Published: 6 May 2019

Abstract: Diarrhoeal disease is still one of the most challenging issues for health in many countries across the Eastern Mediterranean region (EMR), with infectious diarrhoea being an important cause of morbidity and mortality, especially in children under five years of age. However, the understanding of the aetiological spectrum and the burden of enteric pathogens involved in diarrhoeal disease in the EMR is incomplete. Non-typhoidal *Salmonella* (NTS), the focus of this review, is one of the most frequently reported bacterial aetiologies in diarrhoeal disease in the EMR. Strains of NTS with resistance to antimicrobial drugs are increasingly reported in both developed and developing countries. In the EMR, it is now widely accepted that many such resistant strains are zoonotic in origin and acquire their resistance in the food-animal host before onward transmission to humans through the food chain. Here, we review epidemiological and microbiological aspects of diarrhoeal diseases among children in the EMR, with emphasis on the implication and burden of NTS. We collate evidence from studies across the EMR on the zoonotic exposure and antimicrobial resistance in NTS at the interface between human and foods of animal origin. This review adds to our understanding of the global epidemiology of *Salmonella* with emphasis on the current situation in the EMR.

Keywords: Eastern Mediterranean region; non-typhoidal *Salmonella*; zoonoses; child diarrhoea; enteropathogens

---

1. Background

Worldwide, diarrhoeal diseases accounted for 8% of all deaths in children under five years of age in 2016, and this translates to over 1300 young children dying each day or approximately 480,000 children a year [1]. The incidence of diarrhoeal infections among children in the Eastern Mediterranean region (EMR) continues to pose a significant public health challenge in countries across the region. For the purpose of this review, we utilized the regional classification set by the World Health Organization (WHO), and as such, the EMR encompasses the Islamic Republic of Afghanistan (Afghanistan), Djibouti, Somalia, Republic of Yemen (Yemen), Arab Republic of Egypt (Egypt), Islamic Republic of Iran (Iran), Iraq, Jordan, Lebanon, Libya, Morocco, Pakistan, Palestine, Sudan, Syrian Arab Republic (Syria), Tunisia Bahrain, Kuwait, Oman, Qatar, Saudi Arabia, and the United Arab Emirates (UAE). The proportion of paediatric diarrhoea cases has increased over time in several countries in the EMR, as in Iraq—from 14.9% in 1997 to 21.3% in 2000 [2,3]—and Iran, from 10.3% to 19.6% between 2008 and 2010 [4,5]. In Egypt, the incidence of diarrhoea in children has declined from 44% to 23.6% based on reports between 1999 and 2005 [6,7]. Diarrhoea attributed disability-adjusted life years (DALYs) among children under five years of age in the EMR regions were estimated to be 6,058,681

(4,045,101–8,618,353) [8]. Across the 22 countries in the EMR, the highest rates of diarrhoea-attributed mortality among children younger than five years were reported in Somalia, Afghanistan, Djibouti, Iraq, Syria, Yemen, Sudan, Egypt and Tunisia [8,9].

Several countries in the EMR suffer from fragile health care systems, of which Iraq is an example. Several generations of Iraqi children born since the early 1990s have faced adverse conditions negatively impacting their nutrition and health as a result of decades of wars, sanctions and political instability [10]. Diarrhoeal disease is a leading cause of morbidity and mortality among Iraqi children younger than five years [10,11]. The period between 1994 and 1999 witnessed the highest rate of diarrhoea-attributed deaths in Iraq, and in the EMR as a whole, as diarrhoea was a common cause of death in children under five years old; it was responsible for 43.4% of deaths in children aged 2–5 years [11]. Additionally, the 2004 survey of the United Nations International Children's Emergency Fund (UNICEF) in partnership with the Government of Iraq indicated that approximately 90% of children under the age of five years visited hospitals due to diarrhoea [12]. In this former survey, diarrhoea and acute respiratory infection accounted for 70% of childhood deaths, but the fatality rate due to diarrhoeal illnesses was higher than those caused by respiratory infection [12].

2. Epidemiological Aspects of Diarrhoeal Diseases Among Children in the EMR

The majority of diarrhoea infections in children occur during the summer months in countries with a hot and dry climate [13,14]. It has been noted that enteric illnesses in temperate latitudes have a seasonal pattern, with the highest incidence of diseases during the summer months [15]. This is consistent with published evidence of a positive correlation between gastrointestinal infection with enteric pathogens and the increase in ambient temperature [16,17]. Several epidemiological studies confirm the role of age and immune response as important triggers to infectious diarrhoea in children [18,19]. Children below five years of age are significantly more susceptible to diarrhoeal illnesses compared with other age groups [14,18]. In Iraq, a study by Siziya and colleagues (2009) of the prevalence of diarrhoea in 14,676 children less than five years of age revealed that 21.3% of the children had diarrhoea in the two weeks preceding the survey. Based on the aforementioned survey, a history of diarrhoea was positively associated with lower socioeconomic status and a lack of access to clean sources of water [2]. A hospital-based study in Iraq reported a prevalence of diarrhoea in 63.5% of children at three government referral paediatric hospitals in Baghdad [20]. The authors also suggested that this high prevalence rate is likely due to economic collapse, poor sanitation, lack of safe water and inadequate provision of health care [20].

Paediatric diarrhoea has an important financial and productivity impact on the livelihood of families in different countries across the EMR [14,21]. In the United Arab Emirates, Howidi et al. [22] estimated an average cost of $64 for expenses spent dealing with medical care per diarrhoeal episode in children. In Oman, a study by Al Awaidy et al. [23] revealed that the total cost of hospitalization due to diarrhoea (direct medical costs) was estimated at $539 per child for three hospital days, totalling $1.8 million per year for all outpatient and hospital settings in the country.

3. Microbiological Aspects of Diarrhoeal Diseases Among Children in the EMR

The prevalence of enteropathogens in child diarrhoeal illnesses throughout the EMR is difficult to precisely assess due to variations in geographical settings, a lack of harmonization in sampling approaches and study designs and varying laboratory techniques and methods used across different studies, even within the same country [24,25]. Table 1 provides a descriptive summary of various studies reporting the occurrence of major etiological agents responsible for paediatric diarrhoea in different countries across the region. Data on enteric pathogens implicated in diarrhoea among children in the EMR are still limited.

In Iraq, little is known about the causative agents of diarrhoea in children. However, a prospective hospital-based study has shown that *Entamoeba histolytica* is responsible for approximately 85% of diarrhoea infections, and the same study also reported that non-typhoidal *Salmonella* spp. and *Shigella* were isolated from 42% of cases in children under five years of age [26]. In Saudi Arabia, cross-sectional studies have investigated the prevalence of pathogen-induced diarrhoea in faecal samples of children from hospitals and outpatient clinics in different localities. Among the different enteric pathogens found in these studies, rotavirus, *Salmonella* and *Giardia lamblia* were the most prevalent [27–29]. Studies conducted in Bahrain [30], Kuwait [31] and Oman [32] (Table 1) shared some common findings, with rotavirus and adenovirus found to be the major viral causes and *Salmonella* and *Shigella* found to be the most common bacterial causes involved in cases of child diarrhoea [31,32]. The authors also observed that symptoms associated with bacterial gastroenteritis were more severe compared to those of a viral nature. In Qatar, noroviruses have been implicated as the predominant viral pathogen associated with severe diarrhoea in children [33]. Microbiological studies on bacterial diarrhoeal illness among hospitalized children in Pakistan [34], Egypt [35,36], Iran [24], Palestine [37], Djibouti [38] and Somalia [39] reported that the main etiologic agents were *Campylobacter*, *Salmonella*, *E. coli* and *Shigella*. Clinical findings in these studies varied according to the aetiology of diarrhoea; however, abdominal pain, vomiting, fever and dehydration were seen in a majority of cases, and the highest incidence rates were commonly reported in the summer months.

In Libya [40], Sudan [25] and Tunisia [41], the molecular screening of the aetiology of acute diarrhoea among young children has revealed that the major viral agents identified were rotavirus and norovirus, the most frequently diagnosed bacterial pathogens were *Salmonella* spp. and *E. coli* and the most commonly detected parasites were *Giardia lamblia* and *Entamoeba histolytica*. Overall, the above reported studies across different countries in the EMR (Table 1) suggest that knowledge of the aetiology of diarrhoea is important for guiding future epidemiological surveillance and for the implementation of evidence-based public health measures to prevent and control this disease syndrome. *Salmonella* has been featured as one of the leading bacterial causes commonly detected in child diarrhoeal cases across the EMR. In the following section of this review, we will elucidate the state of epidemiological and microbiological features of non-typhoidal *Salmonella* (NTS) implicated in acute paediatric gastroenteritis in children in this region.

Table 1. Distribution of different pathogens from diarrhoeal stool samples among children across countries in the Eastern Mediterranean region.

Country/Locality	Study Period/Month	Population/Age	No. of Stool Samples	Prevalence of Enteropathogens			Reference
				Virus (%)	Bacteria (%)	Parasite (%)	
Bahrain	20	Children < 15 years	805	Rotavirus (13.9) Adenovirus (0.6)	*Salmonella* spp. (5.7) *Shigella* spp. (3.2) *Campylobacter jejuni* (1.6) Enteropathogenic *Escherichia coli* (*E. coli*) (0.5)	ND (not detected)	[30]
Djibouti	1	Children < 16 years	209	ND	Enteroadherent *E. coli* (EAEC) (10.6) Enterotoxigenic *E. coli* (ETEC) (11.0) Enteropathogenic *E. coli* (EPEC) (7.7) *Shigella* spp. (7.7) *Salmonella* spp. (2.9) *Campylobacter jejuni/coli* (3.3) *Aeromonas hydrophila* (3.3)	ND	[38]
Egypt/Alexandria	Not described	Children mean age 9.8 months	880	ND	*Campylobacter* spp. (17.2) *Salmonella* spp. (3) *Shigella* spp. (2)	ND	[35]
Egypt/Fayoum	2	Children < 5 years	356	Rotavirus (17)	Enterotoxigenic *E. coli* (ETEC) (10.8) *Campylobacter* spp. (5.6) *Shigella* spp. (2.0) *Salmonella* spp. (0.6) *Aeromonas hydrophila* (1.1) *Vibrio fluvialis* (0.6)	*Cryptosporidium* (10.7)	[36]
Iran/Tehran	24	Children < 5 years	1078	ND	*Shigella* spp (26.7) *Shiga-like toxin producing E. coli* (STEC) (18.9) Enteroaggregative *E. coli* (EAEC) (16.6) Enteropathogenic *E. coli* (EPEC) (12.6) *Campylobacter* spp. (10.8) *Salmonella* spp. (7.6) Enterotoxigenic *E. coli* (ETEC) (6.8)	ND	[24]
Iraq/Baghdad	Not described	Children < 10 years	1500	ND	*Salmonella* spp. (4.28) *Shigella* spp. (2.14)	*Entamoeba histolytica* (83.58)	[28]

Table 1. *Cont.*

Country/Locality	Study Period/Month	Population/Age	No. of Stool Samples	Prevalence of Enteropathogens			Reference
				Virus (%)	Bacteria (%)	Parasite (%)	
Jordan/Irbid	12	Children < 12 years	265	Rotavirus (32.5)	*Enteropathogenic E. coli* (12.8) *Enteroaggregative E. coli* (10.2) *Enterotoxigenic E. coli* (5.7) *Shigella* spp. (4.9) *Salmonella* spp. (4.5) *Campylobacter jejuni/coli* (1.5) *Enteroinvasive E. coli* (1.5)	*Entamoeba histolytica* (4.9) *Cryptosporidium* spp. (1.5) *Giardia lamblia* (0.8)	[42]
Kuwait	15	Children (not described)	621	Rotavirus (45.0) Adenovirus (4.0)	*Salmonella* spp. (24) *Enterotoxigenic E. coli* (9) *Campylobacter jejuni* (7) *Enteropathogenic E. coli* (7) *Shigella* (4)	ND	[31]
Libya/Tripoli	8	Children < 5 years	239	Norovirus (15.5) Rotavirus (13.4) Adenovirus (7.1) Astrovirus (1.7)	*E. coli* (11.2) *Salmonella* spp. (9.7) *Shigella* spp. (0.8) *Campylobacter* spp. (2.9) *Aeromonas* spp. (4.2) *Cryptosporidium* spp. (2.1)	*Entamoeba histolytica* (0.8) *Giardia lamblia* (1.3)	[40]
Morocco/Rabat	13	Children < 5 years	122	Rotavirus (17.2) Astrovirus (4.9) Hepatitis A (0.8) Norovirus (0.8)	*E. coli* (58.2) *Shigella* spp. (7.4) *Salmonella* spp. (4.1) *Campylobacter* spp. (4.1)	*Giardia intestinalis* (0.8) *Entamoeba histolytica* (0.8)	[43]
Oman/Muscat	24	Children < 5 years	217	Rotavirus (31.0) Adenovirus (4.0)	*E. coli* (10) *Shigella* (7) *Campylobacler* spp. (2) *Salmonella* spp. (2)	*Giardia lamblia* (11), *Entamoeba histolytica* (9)	[32]
Pakistan/Karachi and Rawalpindi	24	Children < 3 years	402	Rotavirus (8.2)	*Enteropathogenic E. coli* (EPEC) (32.8) *Enterotoxigenic E. coli* (ETEC) (14.2) *Shigella* spp. (3.2) *Salmonella* spp. (2)	ND	[34]
Palestine/Gaza	12	Children < 12 years	300	ND	*Enterohemorrhagic E. coli* (8.3) *Shigella* spp. (6.7) *Campylobacter jejuni* (5) *Salmonella* spp. (4) *Yersinia enterocolitica* (2.7) *Aeromonas* spp. (4.7) *Plesiomonas* spp. (1.3)	ND	[37]

Table 1. *Cont.*

Country/Locality	Study Period/Month	Population/Age	No. of Stool Samples	Prevalence of Enteropathogens			Reference
				Virus (%)	Bacteria (%)	Parasite (%)	
Qatar	6	Children (not described)	288	Norovirus (28.5) Rotavirus (10.4) Adenovirus (6.25) Astrovirus (0.30)	*Salmonella* spp. (8) *Escherichia coli* (3) *Shigella* spp. (1.5) *Campylobacter* spp. (1.5)	ND	[33]
Saudi Arabia/Eastern Province	19	Children (not described)	853	Rotavirus (11.5)	*Salmonella* spp. (34) *Shigella* spp. (14.7)	*Entamoeba histolytica* (13.5) *Giardia intestinalis* (10.4)	[27]
Saudi Arabia/Jeddah	12	Children < 5 years	576	Rotavirus (34.6)	*E. coli* (13) *Klebsiella pneumoniae* (4) *Salmonella* spp. (3) *Shigella flexneri* (2.6)	*Giardia lamblia* (3.1) *Entamoeba histolytica* (2.2) *Trichuris trichiura* (0.7) *Hymenolepis nana* (0.7) *Ascaris lumbricoides* (0.7)	[28]
Saudi Arabia/Najran region	9	Children < 5 years	326	Rotavirus (17.2) Adenovirus (3.7) Astrovirus (1.2)	*Salmonella* spp. (8.6) *Shigella* spp. (2.1)	*Giardia lamblia* (0.9) *Entamoeba histolytica* (0.3)	[29]
Somalia Mogadishu	12	Children < 14 years	1667	Rotavirus (25)	*Enterotoxigenic E. coli* (ETEC) (11) *Shigella* spp. (9) *Campylobacter jejuni* (8) *Vibrio cholerae non-O1* (6) *Salmonella* spp. (4) *Aeromonas hydrophila* (9) *Plehnwnas shigelloides* (2)	*Giardia intestinalis* (8) *Entamoeba histolytica* (2)	[39]
Sudan/Khartoum	12	Children < 5 years	437	Rotavirus A (22)	*Enteroaggregative E. coli* (EAEC) (21) *Enteropathogenic E. coli* (EPEC) (14) *Enterotoxigenic E. coli* (ETEC) (9) *Enteroinvasive E. coli* (EIEC) (4) *Shigella sonnei* (3) *Shigella flexneri* (4) *Shigella dysenteriae* (1) *Salmonella typhi* (2) *Salmonella paratyphi C* (1) *Campylobacter jejuni* (3)	*Giardia intestinalis* (11), *Entamoeba histolytica* (5)	[25]
Tunisia	12	Children < 5 years	124	Rotavirus (33.9) Norovirus (8.9)	*Enteroaggregative E. coli* (EAEC) (23.4) *Enteroinvasive E. coli* (EIEC) (12.1) *Enteropathogenic E. coli* (EPEC) (13.7) *Enterotoxigenic E. coli* (ETEC) (21) *Enterohemoragic E. coli* (EHEC) (1.6) *Salmonella* spp. (9.7)	*Entamoeba coli* (1.6) *Cryptosporidies* (1.6) *Giardia lamblia* (0.8) *Blastocystis hominis* (0.8)	[41]

4. Non-Typhoidal *Salmonella* (NTS)—The Pathogen, Exposure and Illness

As of 2012, t more than 2587 serovars of *Salmonella enterica* have been reported from all over the world, and almost all are able to cause illness in animals and humans including gastroenteritis and other acute infections [44]. *Salmonella* spp. are capable of adapting, growing and/or surviving in a various range of environments including temperatures as high as 54 °C or as low as 2 °C, extracellular pH levels below 3.9 and up to 9.5 and salt concentrations up to 4% wv−1 NaCl [45–47]. Such exceptional characteristics can have a significant effect on the survival of *Salmonella* outside of the host organism and in food during processing, preparation, and storage [45,48]. In pure cultures, *Salmonella* spp. are normally inactivated by frozen storage at −22 °C in as few as 5 days [49]; however, freezing does not eliminate the pathogen from contaminated foods [50]. In addition to its survival in extreme conditions, the growth of this pathogen in non-host environments such as natural waters, wastewater sludge, soil and compost has also been reported in several studies [51,52].

There are two major clinical syndromes caused by *Salmonella* infection in human: the first is typhoid and paratyphoid fever, caused by *S.* Typhi and Paratyphi, which are highly adapted to the human host; and the second major clinical syndrome is the gastrointestinal disease caused by a large number of NTS serovars, which are predominantly found in animal reservoirs [53]. The most common mode of NTS infection in human is the ingestion of contaminated food or water [54,55]. Initial symptoms are characterized by an acute onset of fever and chills, nausea and vomiting, abdominal cramping and diarrhoea, and other nonspecific complaints including headache, myalgias, and arthralgias [56,57]. Gastroenteritis caused by NTS is usually self-limiting, lasting for 10 days or less, and may be grossly bloody [54]. *Salmonella* is excreted in faeces after infection, a process that may last for a median of 5 weeks; however, the excretion may be prolonged in young children [58,59]. In rare cases, NTS infections could develop atypical clinical syndromes of variable severity, including bacteraemia, endovascular infection and focal infection [56,60]. In developing countries, children with bacteraemia are more likely to have predisposing conditions, a higher risk for the incidence of meningitis and increased fatality rates compared to adults [58,61].

5. Implication of NTS in Diarrhoeal Illnesses in the EMR

The global burden of NTS gastroenteritis is estimated to be 93.8 million human infections, with 155 thousand deaths and an average incidence rate of 1.14 episodes/100,000 persons [62]. This reflects the enormous burden of the disease in both industrialized and developing countries [63]. For the WHO-defined EMR, the median incidence rate of NTS is 1610 illnesses, with 0.6 deaths per 100,000 persons [64]. The incidence rate of salmonellosis varies substantially between countries across the EMR and is influenced largely by the absence of systematic, harmonized national and regional surveillance and reporting systems. An epidemiological survey in Qatar spanning eight years (2004 to 2012) reported that the incidence rate of reported NTS associated illnesses ranged between 12.3 and 18.1 cases per 100,000 inhabitants, with most reported NTS cases occurring in children under the age of five [65]. In Lebanon, the Department of Epidemiologic Surveillance data from 2001 to 2013 indicated that the annual incidence rate of reported salmonellosis was 13.3 per 100,000 individuals, with an increasing trend of the number of NTS cases between 2009 and 2013 [66]. In Jordan, the reported rate of human salmonellosis is alarmingly higher than what is reported elsewhere across the EMR, with a notification rate of 124 cases per 100,000 persons, as reported in a study from 2003 to 2004 [67]. Significant associations between climatic factors and NST infections have been reported in Iraq [68], Jordan [42], Tunisia [69], Iran [24], Saudi Arabia [70] and Qatar [33]. It has been documented that ambient temperatures contribute directly to *Salmonella* multiplication in foods, water and contaminated environments and thus propagate the likelihood of infection [15,71].

Few studies have been carried out to elucidate the epidemiology of gastrointestinal salmonellosis in the EMR, particularly on children. The prevalence rates of NTS range from as low as 0.2% to as high as 34% [27,39], with the highest reported age-related prevalence usually among children under the age of five (Table 2). Published studies reporting the rates of NTS in the EMR countries are summarised in Table 2. Studies from Iraq (Mosul) [68], Iran (Tehran) [72], Saudi Arabia [27], Kuwait [31], Morocco (Marrakesh) [43] and Yemen [73] reveal a noteworthy high incidence rate (15% to 34%) of NTS (Table 2).

Table 2. Prevalence of *Salmonella* detected in children with acute diarrhoea in the Eastern Mediterranean region.

Country/Locality	Study Period/Month	Population/Age	Source of Cases	Methods	Number of Acute Diarrhoeal Cases	Prevalence of *Salmonella* Infected (%)	Predominate S. Serovar	References
Bahrain	24	Children < 15 years	Hospital admissions	*Salmonella* culture (stool)	805	46 (5.7)	Typhimurium	[30]
Djibouti	1	Children < 16 years	Health centres	*Salmonella* culture (stool)	209	6 (2.9)	NS (not specified)	[38]
Egypt/Aswan	Not described	Children (not described)	Outpatients and inpatients	*Salmonella* culture (stool)	151	11 (7)	NS	[74]
Egypt/Cairo	Not described	Children < 5 years	Hospital admissions	*Salmonella* culture (stool)	356	5 (1.4)	NS	[75]
Egypt/Fayoum	2	Children < 5 years	Hospital admissions	*Salmonella* culture (stool)	356	2 (0.6)		[36]
Iran	36	Children < 5 years	Children hospitals admissions	*Salmonella* culture and PCR	555	42 (7.6)	NS	[24]
Iran/Tehran	108	Children < 14 years	Hospital admissions	*Salmonella* culture (stool)	2487	700 (28.14)	NS	[72]
Iran/Tehran	24	Children < 12 years	Paediatric hospital admissions	*Salmonella* culture (stool)	5900	139 (2.4)	Typhimurium Enteritidis	[76]
Iraq/Mosul	Not described	Children < 7 years	Paediatric hospital admissions	*Salmonella* culture (stool)	111	17 (15)	Typhimurium Worthington	[68]
Iraq/Baghdad	Not described	Children < 10 years	Paediatric hospital admissions	*Salmonella* culture (stool)	420	18 (4.7)	NS	[26]
Jordan/Irbid	12	Children < 12 years	Hospitalized children	*Salmonella* culture (stool)	265	12 (4.5)	NS	[42]
Jordan	48	Children (Not described)	Hospital admissions	*Salmonella* culture (stool)	1400	150 (10.7)	NS	[77]
Kuwait	15	Children (Not described)	Hospitalized children	*Salmonella* culture (stool)	621	149 (24)	NS	[31]
Libya/Zliten	12	Children < 12 years	Hospital admissions	*Salmonella* culture (stool)	169	23 (13.6)	Heidelberg Enteritidis	[78]
Morocco/Marrakesh	12	Children < 15 years	Patient children in household	*Salmonella* culture (stool)	390	127 (32·56)	*Salmonella* group A *Salmonella* group B *Salmonella* group C *Salmonella* group D	[79]
Morocco/Rabat	13	Children < 5 years	Hospital admissions	*Salmonella* culture (stool)	122	5 (4.1)	NS	[43]
Oman/Muscat	24	Children < 5 years	Hospital admissions	*Salmonella* culture (stool)	217	5 (2)	NS	[32]

Table 2. *Cont.*

Country/Locality	Study Period/Month	Population/Age	Source of Cases	Methods	Number of Acute Diarrhoeal Cases	Prevalence of *Salmonella* Infected (%)	Predominate S. Serovar	References
Pakistan/Karachi and Rawalpindi	24	Children < 3 years	Hospital admissions	*Salmonella* culture (stool)	402	8 (2)	NS	[34]
Palestine/Gaza	12	Children < 12 years	Hospital admissions	*Salmonella* culture (stool)	300	12 (4)	NS	[37]
Qatar	5	Children (Not described)	Hospital admissions	*Salmonella* culture (stool)	288	23 (8)	NS	[33]
Saudi Arabia/Yanbu	4	Children (Not described)	Hospital admissions	*Salmonella* culture (stool)	136	15 (11)	Typhimurium Enteritidis Virchow	[80]
Saudi Arabia/Eastern Province	19	Children (Not described)	Hospital admissions	*Salmonella* culture (stool)	853	290 (34)	NS	[27]
Saudi Arabia/South Jeddah	12	Children < 16 years	Hospital admissions	*Salmonella* culture (stool)	367	56 (15.3)	NS	[70]
Somalia/Mogadishu	12	Children < 14 years	Hospital admissions	*Salmonella* culture (stool)	1667	4 (0.2)	NS	[39]
Sudan/Khartoum	12	Children < 5 years	Children in rural area	*Salmonella* culture (stool)	437	3 (0.7)	Enteritidis	[81]
Tunisia/Tunis	48	Children 1–15 years	Paediatric and health centres	*Salmonella* culture (stool)	115	11 (9.5)	Enteritidis Anatum	[69]
Yemen/Thamar	12	Children (Not described)	Hospital admissions	*Salmonella* culture (stool, blood, urine)	460	73 (15.9)	Typhimurium Enteritidis	[73]

Several published studies (Table 2) indicate that the most widely reported serovars associated with acute diarrhoeal disease across the EMR are the *Salmonella* serovar Typhimurium and *Salmonella* serovar Enteritidis [73,76,80]. Similar to the situation in EMR, *S.* Typhimurium followed by *S.* Enteritidis are also the top-ranked serovars involved in human diarrhoeal illnesses across Africa, North America and Oceania (Australia and New Zealand) [82]. In contrast, *S.* Enteritidis is more frequently reported than *S.* Typhimurium in human clinical isolates in Europe, Asia and Latin America [83].

The difference in *Salmonella* prevalence and the diversity of NTS serovars among humans is dynamic in nature, and it is not surprising to capture variations between countries. Such variations might be attributed to several factors impacting NTS levels in food and water, which play a major role in human exposure to infection [84]. Among these factors are climate, food-animal production practices, the level of spread of specific serotypes in environmental reservoirs and the availability of vaccination programs in food animals. Eggs and poultry products have been described as the main vehicles for the transmission of human salmonellosis, accounting for the majority of foodborne outbreaks [84,85].

6. NTS and the Risk of Zoonotic Exposure from Chicken Meat Products

Several studies in the EMR investigated the prevalence of NTS in local and imported chicken meat, as summarised in Table 3. Rates of *Salmonella* contamination vary between countries due to a number of factors including the source and type of sample, slaughterhouse sanitation, the level of cross-contamination at the retail level and the detection methods employed [76,86]. In Iraq, *Salmonella* contamination was reported in 26% of fresh retail chicken meat samples [87,88] and in 39% of raw and frozen chicken carcasses [89]. Studies in several EMR countries have identified low *Salmonella* prevalence rates in chicken meat, such as in Kuwait [90], Tunis/Tunisia [91], Saudi Arabia/Riyadh [92], and Egypt [93,94]. Several studies indicated that *S.* Typhimurium and *S.* Enteritidis were the most prevalent serovars in chicken meat in the EMR [86,93,95,96]. Interestingly, the two most commonly reported *Salmonella* serovars in human diarrhoeal illness are also the two commonly recovered serovars from chicken meat, highlighting the role of chicken meat as an important source of salmonellosis in this region.

Table 3. Prevalence of *Salmonella* detected in chicken meat in the Eastern Mediterranean region.

Country/Locality	Source of Samples	No. Positive/No. Test (% Prevalence)	Methods	Predominant S. Serovars	Reference
Egypt/Dakahlia, Gharbia, Damietta and Kafr el-Sheikh	Chicken meat	14/320 (5)	*Salmonella* culture and PCR	Typhimurium Enteritidis Infantis	[93]
Egypt/Mansoura	Chicken carcasses Chicken drumsticks Chicken gizzards Chicken livers	8/50 (16) 14/50 (28) 16/50 (32) 30/50 (60)	*Salmonella* culture and PCR	Enteritidis Typhimurium Kentucky	[86]
Egypt/Minoufiya and Cairo	Chicken meat	6/100 (6)	*Salmonella* culture and matrix-assisted laser desorption/ionization time-of-flight (MALDI-TOF)	Typhimurium Enteritidis	[94]
Egypt/Zagazig	Chicken meat	7/50 (14)	*Salmonella* culture and PCR	Typhimurium	[97]
Iran/Alborz	Chicken meat Chicken liver Chicken heart Chicken gizzard	58/200 (29) 26/120 (21.6) 17/120 (14.1) 10/120 (8.3)	*Salmonella* culture	Thompson Enteritidis Typhimurium	[87]
Iran/Tehran	Chicken meat	86/190 (45)	*Salmonella* culture	Thompson Haardt Enteritidis	[98]
Iraq/Baghdad	Chicken meat	39/100 (39)	*Salmonella* culture	Infantis Enteritidis Vichow	[89]
Iraq/Mosul	Chicken meat	21/81(26)	*Salmonella* culture	Infantis Zanzibar Anatum	[88]
Jordan/Irbid	Chicken meat	91/104 (87.5)	*Salmonella* culture and PCR	NS	[99]
Jordan/Northern	Chicken product (shawarma)	30/80 (37.5)	*Salmonella* culture and PCR	Paratyphi A Cholerasuis Pullorum	[100]
Jordan/Northern	Chicken meat	200/302 (66)	*Salmonella* culture and PCR	Enteritidis Cholerasuis,	[101]

Table 3. *Cont.*

Country/Locality	Source of Samples	No. Positive/No. Test (% Prevalence)	Methods	Predominant S. Serovars	Reference
Kuwait	Chicken carcasses	11/180 (6.1)	*Salmonella* culture	Enteritidis Infantis	[90]
Libya/Tripoli	Chicken product (burgers)	15/120 (12.9)	*Salmonella* culture	NS	[102]
Morocco/Tetouan	Chicken meat	18/86 (20.9)	*Salmonella* culture	Kentucky Typhimurium, Enteritidis Agona	[95]
Pakistan/Hyderabad	Chicken meat	38/100 (38)	*Salmonella* culture	Enteritidis Typhimurium.	[103]
Pakistan/Karachi	Chicken carcasses	78/160 (48.75)	*Salmonella* culture	Enteritidis Typhimurium	[96]
Saudi Arabia/Riyadh	Chicken liver Chicken heart Chicken spleen	209/3284 (6.3) 107/2315 (4.6) 45/899 (5.0)	*Salmonella* culture	Enteritidis Virchow	[92]
Sudan/Khartoum	Chicken meat	35/193 (18.13)	*Salmonella* culture	Stanleyville Kentucky Virchow Hadar Typhimurium	[104]
Syria	Chicken meat	32/100 (32)	*Salmonella* culture	NS	[105]
Tunis	Chicken meat	29/60 (48.3)	*Salmonella* culture and PCR	Typhimurium Zanzibar Orion	[106]
Tunis/Tunisia	Chicken meat	29/433 (6.7)	*Salmonella* culture and PCR	Kentucky Anatum Zanzibar	[91]
United Arab Emirates/Dubai	Chicken meat	28/60 (46.67)	*Salmonella* culture	NS	[107]

To study the zoonotic transmission of NTS at the human–animal–food interface, it is important to employ the advances in molecular epidemiology tools. Pulsed-field gel electrophoresis (PFGE) and multiple-locus variable-number tandem repeat analysis (MLVA) have been widely considered to be the gold standard molecular subtyping and fingerprinting methods for tracking the source of *Salmonella* contamination [108,109]. However, in recent years, Whole Genome Sequencing (WGS) has become a powerful tool in elucidating transmission pathways [110]. For *Salmonella*, WGS provides a massive amount of data for research purposes along with the rapid acquisition of multilocus sequence types (MLST), serotypes and antimicrobial resistance gene data [111–113]. Sequence-based typing can also be used to obtain basic biological insights to explain the associations between isolates, thus providing added value to the source attribution [112–114]. In addition to its high discrimination ability, WGS could provide additional data about virulence determinants and genome evolution, and such results can be easily shared and communicated [115–117].

7. Antimicrobial Resistance in NTS at the Interface between Human and Food of Animal Origin in the EMR

Antimicrobial therapy is recommended in severe cases and/or cases of prolonged enteritis, meningitis, septicaemia and extra-intestinal complications associated with salmonellosis [53,58,60]. Antimicrobial resistance in NTS has increased in recent years worldwide, due to the widespread use of antimicrobial drugs in human and veterinary sectors, and poses an on-going threat to global public health [118–121]. The incidence of resistance to traditional antibiotics (e.g., ampicillin, tetracycline and streptomycin) is evident to be high in *Salmonella* isolates from foods of animal origin, especially poultry, in EMR countries [86,87,98–120]. This finding is highly concerning from a public health perspective, as many of these traditional (1st generation) antibiotics are still widely prescribed to treat diarrhoea in children and adults due to their low cost and availability in developing countries, including countries of the EMR [121,122]. Similar patterns of high resistance to these traditional antibiotics are also evident in *Salmonella* isolated from human enteric illnesses, especially in Iraq [68], Kuwait [31], Saudi Arabia [123], Jordan [77], Iran [76], Oman [124] and Libya [40].

Studies in Saudi Arabia [123], and Kuwait [125] reported frequent resistance to chloramphenicol in NTS isolated from childhood diarrhoea (although it is not approved for human use). Resistance to chloramphenicol in *Salmonella* is facilitated by type A chloramphenicol acetyltransferase genes (*catA1* and *catA2*) or by cassette-borne type B chloramphenicol acetyltransferase genes (*catB2*, *catB3* or *catB8*) [126,127]. Furthermore, two new chloramphenicol/florfenicol exporter genes, *cmlA9* and *floR*, have recently been identified for phenicol resistance genes in *Salmonella* isolates [121–128].

Resistance to nalidixic acid (NA) in *Salmonella* isolates from paediatric cases with enteritis was as high as 84.2% in a study in Libya [40] and was detected at a rate of 42.3% in a study in Iran [72]. There is an alarming concern over the increase in the resistance of NTS to ciprofloxacin and extended-spectrum cephalosporins [118,129], given the critical clinical relevance of these antimicrobials. Chromosomal mutations in the genes encoding topoisomerase II, *gyrA* and *gyrB*, and/or topoisomerase IV, *parC* and *parE*, accounting for resistance to quinolones/fluoroquinolones, are known to occur in *Salmonella* isolates [130]. More recently, various plasmid-mediated quinolone resistance (PMQR) genes including genes *qnrD*, *qnrA*, *qnrB* and *qnrS* variants, all of which code for DNA topoisomerase protecting proteins, as well as the genes *qepA* and *qepAB* coding for a quinolone-specific efflux pump, and the acetyltransferase *aac(60)-Ib-cr*, have been identified in *Salmonella* isolates [131–133]. In some EMR countries, the increase in resistance is rapid and considerable; in Libya, a study reported that 63.1% of human *Salmonella* isolates were ciprofloxacin-resistant [40].

Resistance to β-lactam antibiotics (penicillins, cephalosporins, and carbapenems) in *Salmonella* is mediated by a wide range of β-lactamase enzymes [134]. To date, genes coding for 13 different types of β-lactamases have been known in *Salmonella*. Among them, $bla_{AAC}/bla_{DHA}/bla_{CMY}/bla_{TEM}$ genes are of particular importance as the first representative encoding of cephalosporinases that hydrolyse most β-lactamase except carbapenems [129]. Extended-spectrum cephalosporins are the antimicrobials of choice for invasive *Salmonella* treatment, especially in children where treatment with fluoroquinolones is not recommended [130]. A study by Rotimi et al. [135] observed resistance to cefotaxime and ceftriaxone among *Salmonella* spp. isolated from stool samples of patients with acute diarrhoea and septicaemia during 2003–2005 in Kuwait and United Arab of Emirates.

8. Conclusions

This review consolidates recent updates on the spectrum of enteric pathogens in the EMR region, with special emphasis on NTS. Among bacterial pathogens, NTS infections continue to pose a distressing public health concern, notably in children under five years old. The emergence of antimicrobial resistance in *Salmonella* strains present a great challenge at the human–food–environment interface in terms of the effective treatment of the infections caused by these strains. The EMR spans different countries with varying and evolving socio-economic statuses. Several countries in the EMR, notably Iraq, Yemen and Syria, are experiencing similar challenges as a result of fragile political situations and insecurity as well as sanitation, food safety, and food security issues and an influx of refugees. The public health system in several countries in the EMR is struggling to respond to the evolving burden of enteric illnesses due to the lack of surveillance of important enteric pathogens, such as *Salmonella*, at hospital, household and food chain levels and would benefit from a multi-dimensional research approach encompassing these levels. Bacterial infections and their antimicrobial resistance profiles should be monitored more closely across the EMR, especially in vulnerable groups such as children less than five years old. Studies focusing on investigating epidemiological and microbiological aspects of infectious diarrhoea in underprivileged communities/regions at the national level should be prioritized in future research.

Author Contributions: Conceptualization, I.H.; methodology, A.H.; writing—review & editing, A.H., M.O., S.A., I.H.

Funding: This research received no external funding.

Conflicts of Interest: The authors declare no conflict of interest.

References

1. United Nations International Children's Emergency Fund. Diarrhoeal Diseases. UNICEF Data: Monitoring the Situation of Children and Women. 2018. Available online: https://data.unicef.org/topic/child-health/diarrhoeal-disease/ (accessed on 5 June 2018).
2. Siziya, S.; Muula, A.S.; Rudatsikira, E. Diarrhoea and acute respiratory infections prevalence and risk factors among under-five children in Iraq in 2000. *Ital. J. Pediatr.* **2009**, *35*, 8. [CrossRef]
3. Tawfeek, H.I.; Najim, N.H.; Al-Mashikhi, S. Studies on diarrhoeal illness among hospitalized children under 5 years of age in Baghdad during 1990–1997. *East. Mediterr. Health J.* **2002**, *8*, 181–188.
4. Kermani, N.A.; Jafari, F.; Mojarad, H.N.; Zali, M.R. Prevalence and associated factors of persistent diarrhoea in Iranian children admitted to a paediatric hospital. *East. Mediterr. Health J.* **2010**, *16*, 831–836. [CrossRef] [PubMed]
5. Kolahi, A.A.; Nabavi, M.; Sohrabi, M.R. Epidemiology of acute diarrheal diseases among children under 5 years of age in Tehran, Iran. *Iran. J. Clin. Infect. Dis.* **2008**, *3*, 193–198.
6. Deghedi, B.; Mahdy, N.H. Assessment of health and nutritional status of infants in relation to breast feeding practices in Karmouz area, Alexandria. *J. Egypt. Public Health Assoc.* **1999**, *74*, 567–600. [PubMed]
7. El-Gilany, A.H.; Hammad, S. Epidemiology of diarrhoeal diseases among children under age 5 years in Dakahlia, Egypt. *East. Mediterr. Health J.* **2005**, *11*, 762–775.
8. GBD 2015 Eastern Mediterranean Region Diarrhea Collaborators. Burden of diarrhea in the Eastern Mediterranean Region, 1990–2015: Findings from the Global Burden of Disease 2015 study. *Int. J. Public Health* **2018**, *63* (Suppl. 1), 109–121. [CrossRef]
9. GBD Diarrhoeal Diseases Collaborators. Estimates of global, regional, and national morbidity, mortality, and aetiologies of diarrhoeal diseases: A systematic analysis for the Global Burden of Disease Study 2015. *Lancet Infect. Dis.* **2017**, *17*, 909–948. [CrossRef]

10. Ahmed, S.; Klena, J.; Albana, A.; Alhamdani, F.; Oskoff, J.; Soliman, M.; Heylen, E.; Teleb, N.; Husain, T.; Matthijnssens, J. Characterization of human rotaviruses circulating in Iraq in 2008: Atypical G8 and high prevalence of P [6] strains. *Infect. Genet. Evol.* **2013**, *16*, 212–217. [CrossRef]

11. Awqati, N.A.; Ali, M.M.; Al-Ward, N.J.; Majeed, F.A.; Salman, K.; Al-Alak, M.; Al-Gasseer, N. Causes and differentials of childhood mortality in Iraq. *BMC Pediatr.* **2009**, *9*, 40. [CrossRef]

12. United Nations International Children's Emergency Fund. Evaluation of UNICEF Emergency Preparedness and Early Response in Iraq (September 2001–June 2003). 2004. Available online: https://www.unicef.org/evaldatabase/files/Iraq_evaluation_FINAL_2006.pdf (accessed on 6 October 2004).

13. Musengimana, G.; Mukinda, F.K.; Machekano, R.; Mahomed, H. Temperature Variability and Occurrence of Diarrhoea in Children under Five-Years-Old in Cape Town Metropolitan Sub-Districts. *Int. J. Environ. Res. Public Health* **2016**, *13*, 859. [CrossRef] [PubMed]

14. Khoury, H.; Ogilvie, I.; El Khoury, A.C.; Duan, Y.; Goetghebeur, M.M. Burden of rotavirus gastroenteritis in the Middle Eastern and North African pediatric population. *BMC Infect. Dis.* **2011**, *11*, 9. [CrossRef] [PubMed]

15. Akil, L.; Ahmad, H.A.; Reddy, R.S. Effects of climate change on *Salmonella* infections. *Foodborne Pathog. Dis.* **2014**, *11*, 974–980. [CrossRef] [PubMed]

16. Mellor, J.E.; Levy, K.; Zimmerman, J.; Elliott, M.; Bartram, J.; Carlton, E.; Clasen, T.; Dillingham, R.; Eisenberg, J.; Guerrant, R.; et al. Planning for climate change: The need for mechanistic systems-based approaches to study climate change impacts on diarrheal diseases. *Sci. Total Environ.* **2016**, *548–549*, 82–90. [CrossRef] [PubMed]

17. Checkley, W.; Epstein, L.D.; Gilman, R.H.; Figueroa, D.; Cama, R.I.; Patz, J.A.; Black, R.E. Effect of El Nino and ambient temperature on hospital admissions for diarrhoeal diseases in Peruvian children. *Lancet* **2000**, *355*, 442–450. [CrossRef]

18. Gizaw, Z.; Woldu, W.; Bitew, B.D. Child feeding practices and diarrheal disease among children less than two years of age of the nomadic people in Hadaleala District, Afar Region, Northeast Ethiopia. *Int. Breastfeed. J.* **2017**, *12*, 24. [CrossRef] [PubMed]

19. Mihrete, T.S.; Getahun Asres Alemie, G.A.; Teferra, A.S. Determinants of childhood diarrhea among underfive children in Benishangul Gumuz Regional State, North West Ethiopia. *BMC Pediatr.* **2014**, *14*, 102.

20. Alaa, H.; Shah, S.A.; Khan, A.R. Prevalence of diarrhoea and its associated factors in children under five years of age in Baghdad, Iraq. *Open J. Prev. Med.* **2014**, *4*, 17–21. [CrossRef]

21. Sepanlou, S.G.; Malekzadeh, F.; Delavari, F.; Naghavi, M.; Forouzanfar, M.H.; Moradi-Lakeh, M.; Malekzadeh, R.; Poustchi, H.; Pourshams, A. Burden of Gastrointestinal and Liver Diseases in Middle East and North Africa: Results of Global Burden of Diseases Study from 1990 to 2010. *Middle East J. Dig. Dis.* **2015**, *7*, 201–215.

22. Howidi, M.; Al Kaabi, N.; El Khoury, A.C.; Brandtmüller, A.; Nagy, L.; Richer, E.; Haddadin, W.; Miqdady, M.S. Burden of acute gastroenteritis among children younger than 5 years of age-a survey among parents in the United Arab Emirates. *BMC Pediatr.* **2012**, *12*, 74. [CrossRef]

23. Al Awaidy, S.A.; Bawikar, S.; Al Busaidy, S.; Baqiani, S.; Al Abedani, I.; Varghese, R.; Abdoan, H.S.; Al Abdoon, H.; Bhatnagar, S.; Al Hasini, K.S. Considerations for introduction of a rotavirus vaccine in Oman: Rotavirus disease and economic burden. *J. Infect. Dis.* **2009**, *200* (Suppl. 1), S248–S253. [CrossRef]

24. Jafari, F.; Garcia-Gil, L.J.; Salmanzadeh-Ahrabi, S.; Shokrzadeh, L.; Aslani, M.M.; Pourhoseingholi, M.A.; Derakhshan, F.; Zali, M.R. Diagnosis and prevalence of enteropathogenic bacteria in children less than 5 years of age with acute diarrhea in Tehran children's hospitals. *J. Infect.* **2009**, *58*, 21–27. [CrossRef]

25. Saeed, A.; Abd, H.; Sandstrom, G. Microbial aetiology of acute diarrhea in children under five years of age in Khartoum, Sudan. *J. Med. Microbiol.* **2015**, *64*, 432–437. [CrossRef]

26. Al-Kubaisy, W.; Al Badre, A.; Al-Naggar, R.A.; Shamsidah, N.N. Epidemiological study of bloody diarrhoea among children in Baghdad, Iraq. *Int. Arch. Med.* **2015**, *8*. [CrossRef]

27. Al-Freihi, H.; Twum-Danso, K.; Sohaibani, M.; Bella, H.; el-Mouzan, M.; Sama, K. The microbiology of acute diarrhoeal disease in the eastern province of Saudi Arabia. *East Afr. Med. J.* **1993**, *70*, 267–269.

28. El-Sheikh, S.M.; el-Assouli, S.M. Prevalence of viral, bacterial and parasitic enteropathogens among young children with acute diarrhoea in Jeddah, Saudi Arabia. *J. Health Popul. Nutr.* **2001**, *19*, 25–30.

29. Al Ayed, M.S.; Asaad, A.; Mahdi, A.; Qureshi, M. Aetiology of acute gastroenteritis in children in Najran region, Saudi Arabia. *J. Health Spec.* **2013**, *1*, 84–89. [CrossRef]

30. Ismaeel, A.Y.; Jamsheer, A.E.; Yousif, A.Q.; Al-Otaibi, M.A.; Botta, G.A. Causative pathogens of severe diarrhea in children. *Saudi Med. J.* **2002**, *23*, 1064–1069.

31. Sethi, S.K.; Khuffash, F.A.; al-Nakib, W. Microbial etiology of acute gastroenteritis in hospitalized children in Kuwait. *Pediatr. Infect. Dis. J.* **1989**, *8*, 593–597. [CrossRef]

32. Aithala, G.; Al Dhahry, S.H.; Saha, A.; Elbualy, M.S. Epidemiological and clinical features of rotavirus gastroenteritis in Oman. *J. Trop. Pediatr.* **1996**, *42*, 54–57. [CrossRef]

33. Al-Thani, A.; Baris, M.; Al-Lawati, N.; Al-Dhahry, S. Characterising the aetiology of severe acute gastroenteritis among patients visiting a hospital in Qatar using real-time polymerase chain reaction. *BMC Infect. Dis.* **2013**, *13*, 329. [CrossRef] [PubMed]

34. Mubashir, M.; Khan, A.; Baqai, R.; Iqbal, J.; Ghafoor, A.; Zuberi, S.; Burney, M.I. Causative agents of acute diarrhoea in the first 3 years of life: Hospital-based study. *J. Gastroenterol. Hepatol.* **1990**, *5*, 264–270. [PubMed]

35. Pazzaglia, G.; Bourgeois, A.L.; Mourad, A.S.; Iqbal, J.; Ghafoor, A.; Zuberi, S.; Burney, M.I. *Campylobacter* diarrhea in Alexandria, Egypt. *J. Egypt. Public Health Assoc.* **1995**, *70*, 229–241. [PubMed]

36. El-Mohamady, H.; Abdel-Messih, I.A.; Youssef, F.G.; Said, M.; Farag, H.; Shaheen, H.I.; Rockabrand, D.M.; Luby, S.B.; Hajjeh, R.; Sanders, J.W.; et al. Enteric pathogens associated with diarrhea in children in Fayoum, Egypt. *Diagn. Microbiol. Infect. Dis.* **2006**, *56*, 1–5. [CrossRef] [PubMed]

37. Al Jarousha, A.M.; El Jarou, M.A.; El Qouqa, I.A. Bacterial enteropathogens and risk factors associated with childhood diarrhea. *Indian J. Pediatr.* **2011**, *78*, 165–170. [CrossRef] [PubMed]

38. Mikhail, I.A.; Fox, E.; Haberberger, R.L.; Ahmed, M.H.; Abbatte, E.A. Epidemiology of bacterial pathogens associated with infectious diarrhea in Djibouti. *J. Clin. Microbiol.* **1990**, *28*, 956–961.

39. Casalino, M.; Yusuf, M.W.; Nicoletti, M.; Bazzicalupo, P.; Coppo, A.; Colonna, B.; Cappelli, C.; Bianchini, C.; Falbo, V.; Ahmed, H.J.; et al. A two-year study of enteric infections associated with diarrhoeal diseases in children in urban Somalia. *Trans. R. Soc. Trop. Med. Hyg.* **1988**, *82*, 637–641. [CrossRef]

40. Rahouma, A.; Klena, J.D.; Krema, Z.; Abobker, A.A.; Treesh, K.; Franka, E.; Abusnena, O.; Shaheen, H.I.; El Mohammady, H.; Abudher, A.; et al. Enteric pathogens associated with childhood diarrhea in Tripoli-Libya. *Am. J. Trop. Med. Hyg.* **2011**, *84*, 886–891. [CrossRef]

41. Nejma, I.B.S.B.; Zaafrane, M.H.; Hassine, F.; Sdiri-Loulizi, K.; Ben Said, M.; Aouni, M.; Mzoughi, R. Etiology of Acute Diarrhea in Tunisian Children with Emphasis on Diarrheagenic Escherichia coli: Prevalence and Identification of E. coli Virulence Markers. *Iran. J. Public Health* **2014**, *43*, 947–960.

42. Youssef, M.; Shurman, A.; Bougnoux, M.; Rawashdeh, M.; Bretagne, S.; Strockbine, N. Bacterial, viral and parasitic enteric pathogens associated with acute diarrhea in hospitalized children from northern Jordan. *FEMS Immunol. Med. Microbiol.* **2000**, *28*, 257–263. [CrossRef]

43. Benmessaoud, R.; Jroundi, I.; Nezha, M.; Moraleda, C.; Tligui, H.; Seffar, M.; Alvarez-Martínez, M.J.; Pons, M.J.; Chaacho, S.; Hayes, E.B.; et al. Aetiology, epidemiology and clinical characteristics of acute moderate-to-severe diarrhoea in children under 5 years of age hospitalized in a referral paediatric hospital in Rabat, Morocco. *J. Med. Microbiol.* **2015**, *64*, 84–92. [CrossRef] [PubMed]

44. Achtman, M.; Wain, J.; Weill, F.X.; Nair, S.; Zhou, Z.; Sangal, V.; Krauland, M.G.; Hale, J.L.; Harbottle, H.; Uesbeck, A.; et al. Multilocus sequence typing as a replacement for serotyping in *Salmonella* enterica. *PLoS Pathog.* **2012**, *8*, e1002776. [CrossRef] [PubMed]

45. Spector, M.P.; Kenyon, W.J. Resistance and survival strategies of *Salmonella enterica* to environmental stresses. *Food Res. Int.* **2012**, *45*, 455–481. [CrossRef]

46. Xu, H.; Lee, H.Y.; Ahn, J. Cross-protective effect of acid-adapted *Salmonella enterica* on resistance to lethal acid and cold stress conditions. *Lett. Appl. Microbiol.* **2008**, *47*, 290–297. [CrossRef] [PubMed]

47. Jarvis, N.A.; O'Bryan, C.A.; Dawoud, T.M.; Park, S.H.; Kwon, Y.M.; Crandall, G.P.; Ricke, S.C. An overview of *Salmonella* thermal destruction during food processing and preparation. *Food Control* **2016**, *68*, 280–290. [CrossRef]

48. Yang, Y.; Khoo, W.J.; Zheng, Q.; Chung, H.J.; Yuk, H.G. Growth temperature alters *Salmonella* Enteritidis heat/acid resistance, membrane lipid composition and stress/virulence related gene expression. *Int. J. Food Microbiol.* **2014**, *172*, 102–109. [CrossRef] [PubMed]

49. Manios, S.G.; Skandamis, P.N. Effect of frozen storage, different thawing methods and cooking processes on the survival of *Salmonella* spp. and Escherichia coli O157:H7 in commercially shaped beef patties. *Meat Sci.* **2015**, *101*, 25–32. [CrossRef]

50. Simpson Beauchamp, C.; Byelashov, O.A.; Geornaras, I.; Kendall, P.A.; Scanga, J.A.; Belk, K.E.; Smith, G.C.; Sofos, J.N. Fate of *Listeria monocytogenes* during freezing, thawing and home storage of frankfurters. *Food Microbiol.* **2010**, *27*, 144–149. [CrossRef]

51. Islam, M.; Morgan, J.; Doyle, M.P.; Phatak, S.C.; Millner, P.; Jiang, X. Fate of *Salmonella enterica* serovar Typhimurium on carrots and radishes grown in fields treated with contaminated manure composts or irrigation water. *Appl. Environ. Microbiol.* **2004**, *70*, 2497–2502. [CrossRef]

52. Levantesi, C.; Bonadonna, L.; Briancesco, R.; Grohmann, E.; Toze, S.; Tandoi, V. *Salmonella* in surface and drinking water: Occurrence and water-mediated transmission. *Food Res. Int.* **2012**, *45*, 587–602. [CrossRef]

53. Gordon, M.A. *Salmonella* infections in immunocompromised adults. *J. Infect.* **2008**, *56*, 413–422. [CrossRef] [PubMed]

54. Eng, S.-K.; Pusparajah, P.; Ab Mutalib, N.-S.; Ser, H.-L.; Chan, K.-G.; Lee, L.-H. *Salmonella*: A review on pathogenesis, epidemiology and antibiotic resistance. *Front. Life Sci.* **2015**, *8*, 284–293. [CrossRef]

55. Fabrega, A.; Vila, J. *Salmonella enterica* serovar Typhimurium skills to succeed in the host: Virulence and regulation. *Clin. Microbiol. Rev.* **2013**, *26*, 308–341. [CrossRef] [PubMed]

56. Hohmann, E.L. Nontyphoidal salmonellosis. *Clin. Infect. Dis.* **2001**, *32*, 263–269. [PubMed]

57. Jones, T.F.; Ingram, L.A.; Cieslak, P.R.; Vugia, D.J.; Tobin-D'Angelo, M.; Hurd, S.; Medus, C.; Cronquist, A.; Angulo, F.J. Salmonellosis outcomes differ substantially by serotype. *J. Infect. Dis.* **2008**, *198*, 109–114. [CrossRef]

58. Chen, H.M.; Wang, Y.; Su, L.H.; Chiu, C.H. Nontyphoid *Salmonella* infection: Microbiology, clinical features, and antimicrobial therapy. *Pediatr. Neonatol.* **2013**, *54*, 147–152. [CrossRef] [PubMed]

59. Crump, J.A.; Sjolund-Karlsson, M.; Gordon, M.A.; Parry, C.M. Epidemiology, clinical presentation, laboratory diagnosis, antimicrobial resistance, and antimicrobial management of invasive *Salmonella* infections. *Clin. Microbiol. Rev.* **2015**, *28*, 901–937. [CrossRef] [PubMed]

60. Sanchez-Vargas, F.M.; Abu-El-Haija, M.A.; Gomez-Duarte, O.G. *Salmonella* infections: An update on epidemiology, management, and prevention. *Travel Med. Infect. Dis.* **2011**, *9*, 263–277. [CrossRef]

61. Sirinavin, S.; Chiemchanya, S.; Vorachit, M. Systemic nontyphoidal *Salmonella* infection in normal infants in Thailand. *Pediatr. Infect. Dis. J.* **2001**, *20*, 581–587. [CrossRef]

62. Majowicz, S.E.; Musto, J.; Scallan, E.; Angulo, F.J.; Kirk, M.; O'Brien, S.J.; Jones, T.F.; Fazil, A.; Hoekstra, R.M. International Collaboration on Enteric Disease 'Burden of Illness' Studies. The global burden of nontyphoidal *Salmonella* gastroenteritis. *Clin. Infect. Dis.* **2010**, *50*, 882–889. [CrossRef]

63. WHO. Salmonella (Non-Typhoidal). Available online: https://www.who.int/en/news-room/fact-sheets/detail/salmonella-(non-typhoidal) (accessed on 5 February 2018).

64. Kirk, M.D.; Pires, S.M.; Black, R.E.; Caipo, M.; Crump, J.A.; Devleesschauwer, B.; Döpfer, D.; Fazil, A.; Fischer-Walker, C.L.; Hald, T.; et al. World Health Organization Estimates of the Global and Regional Disease Burden of 22 Foodborne Bacterial, Protozoal, and Viral Diseases, 2010: A Data Synthesis. *PLoS Med.* **2015**, *12*, e1001921.

65. Farag, E.; Garcell, H.G.; Ganesan, N.; Ahmed, S.N.; Al-Hajri, M.; Al Thani, S.M.; Al-Marri, S.A.; Ibrahim, E.; Al-Romaihi, H.E. A retrospective epidemiological study on the incidence of salmonellosis in the State of Qatar during 2004–2012. *Qatar Med. J.* **2016**, *2016*, 3.

66. Malaeb, M.; Bizri, A.R.; Ghosn, N.; Berry, A.; Musharrafieh, U. *Salmonella* burden in Lebanon. *Epidemiol. Infect.* **2016**, *144*, 1761–1769. [CrossRef]

67. Gargouri, N.; Walke, H.; Belbeisi, A.; Hadadin, A.; Salah, S.; Ellis, A.; Braam, H.P.; Angulo, F.J. Estimated burden of human *Salmonella*, *Shigella*, and Brucella infections in Jordan, 2003–2004. *Foodborne Pathog. Dis.* **2009**, *6*, 481–486. [CrossRef] [PubMed]

68. al-Rajab, W.J.; Abdullah, B.A.; Shareef, A.Y. *Salmonella* responsible for infantile gastroenteritis in Mosul, Iraq. *J. Trop. Med. Hyg.* **1988**, *91*, 315–318. [PubMed]

69. Al-Gallas, N.; Bahri, O.; Bouratbeen, A.; Ben Haasen, A.; Ben Aissa, R. Etiology of acute diarrhea in children and adults in Tunis, Tunisia, with emphasis on diarrheagenic *Escherichia coli*: Prevalence, phenotyping, and molecular epidemiology. *Am. J. Trop. Med. Hyg.* **2007**, *77*, 571–582. [CrossRef] [PubMed]

70. Hegazi, M.A.; Patel, T.A.; El-Deek, B.S. Prevalence and characters of *Entamoeba histolytica* infection in Saudi infants and children admitted with diarrhea at 2 main hospitals at South Jeddah: A re-emerging serious infection with unusual presentation. *Braz. J. Infect. Dis.* **2013**, *17*, 32–40. [CrossRef] [PubMed]

71. Ravel, A.; Smolina, E.; Sargeant, J.M.; Cook, A.; Marshall, B.; Fleury, M.D.; Pollari, F. Seasonality in human *salmonellosis*: Assessment of human activities and chicken contamination as driving factors. *Foodborne Pathog. Dis.* **2010**, *7*, 785–794. [CrossRef] [PubMed]

72. Ashtiani, M.T.; Monajemzadeh, M.; Kashi, L. Trends in antimicrobial resistance of fecal *Shigella* and *Salmonella* isolates in Tehran, Iran. *Indian J. Pathol. Microbiol.* **2009**, *52*, 52–55.

73. Taha, R.R.; Alghalibi, S.M.; Saeedsaleh, M.G. *Salmonella* spp. in patients suffering from enteric fever and food poisoning in Thamar city, Yemen. *East. Mediterr. Health J.* **2013**, *19*, 88–93. [CrossRef]

74. Mikhail, I.A.; Hyams, K.C.; Podgore, J.K.; Haberberger, R.L.; Boghdadi, A.M.; Mansour, N.S.; Woody, J.N. Microbiologic and clinical study of acute diarrhea in children in Aswan, Egypt. *Scand. J. Infect. Dis.* **1989**, *21*, 59–65. [CrossRef] [PubMed]

75. El-Shabrawi, M.; Salem, M.; Abou-Zekri, M.; El-Naghi, S.; Hassanin, F.; El-Adly, T.; El-Shamy, A. The burden of different pathogens in acute diarrhoeal episodes among a cohort of Egyptian children less than five years old. *Prz. Gastroenterol.* **2015**, *10*, 173–180. [CrossRef]

76. Ranjbar, R.; Giammanco, G.M.; Farshad, S.; Owlia, P.; Aleo, A.; Mammina, C. Serotypes, antibiotic resistance, and class 1 integrons in *Salmonella* isolates from pediatric cases of enteritis in Tehran, Iran. *Foodborne Pathog. Dis.* **2011**, *8*, 547–553. [CrossRef] [PubMed]

77. Battikhi, M.N. Epidemiological study on Jordanian patients suffering from diarrhoea. *New Microbiol.* **2002**, *25*, 405–412. [PubMed]

78. Ali, M.B.; Ghenghesh, K.S.; Aissa, R.B.; Abuhelfaia, A.; Dufani, M. Etiology of childhood diarrhea in Zliten, Libya. *Saudi Med. J.* **2005**, *26*, 1759–1765. [PubMed]

79. Aiat Melloul, A.; Hassani, L. *Salmonella* infection in children from the wastewater-spreading zone of Marrakesh city (Morocco). *J. Appl. Microbiol.* **1999**, *87*, 536–539. [CrossRef]

80. Ramadan, F.; Unni, A.G.; Hablas, R.; Rizk, M.S. *Salmonella*-induced enteritis. Clinical, serotypes and treatment. *J. Egypt. Public Health Assoc.* **1992**, *67*, 357–367.

81. Adam, M.A.; Wang, J.; Enan, K.A.; Shen, H.; Wang, H.; El Hussein, A.R.; Musa, A.B.; Khidir, I.M.; Ma, X. Molecular Survey of Viral and Bacterial Causes of Childhood Diarrhea in Khartoum State, Sudan. *Front. Microbiol.* **2018**, *9*, 112. [CrossRef]

82. Hendriksen, R.S.; Vieira, A.R.; Karlsmose, S.; Lo Fo Wong, D.M.; Jensen, A.B.; Wegener, H.C.; Aarestrup, F.M. Global monitoring of *Salmonella* serovar distribution from the World Health Organization Global Foodborne Infections Network Country Data Bank: Results of quality assured laboratories from 2001 to 2007. *Foodborne Pathog. Dis.* **2011**, *8*, 887–900. [CrossRef]

83. Galanis, E.; Wong, D.M.L.F.; Patrick, M.E.; Binsztein, N.; Cieslik, A.; Chalermchikit, T.; Aidara-Kane, A.; Ellis, A.; Angulo, F.J.; Wegener, H.C.; et al. Web-based surveillance and global *Salmonella* distribution, 2000–2002. *Emerg. Infect. Dis.* **2006**, *12*, 381–388. [CrossRef]

84. Kasimoglu Dogru, A.; Ayaz, N.D.; Gencay, Y.E. Serotype identification and antimicrobial resistance profiles of *Salmonella* spp. isolated from chicken carcasses. *Trop. Anim. Health Prod.* **2010**, *42*, 893–897. [CrossRef]

85. Ayers, L.T.; Williams, I.T.; Gray, S.; Griffin, P.M.; Hall, A.J. Surveillance for Foodborne Disease Outbreaks—United States, 2006. *Morb. Mortal. Wkly. Rep.* **2009**, *58*, 609–615. Available online: https://www.cdc.gov/mmwr/preview/mmwrhtml/mm5822a1.htm (accessed on 12 June 2009).

86. Abd-Elghany, S.M.; Sallam, K.I.; Abd-Elkhalek, A.; Tamura, T. Occurrence, genetic characterization and antimicrobial resistance of *Salmonella* isolated from chicken meat and giblets. *Epidemiol. Infect.* **2015**, *143*, 997–1003. [CrossRef]

87. Sodagari, H.R.; Mashak, Z.; Ghadimianazar, A. Prevalence and antimicrobial resistance of *Salmonella* serotypes isolated from retail chicken meat and giblets in Iran. *J. Infect. Dev. Ctries.* **2015**, *9*, 463–469. [CrossRef]

88. Al-Rajab, W.J.; Al-Chalabi, K.A.; Sulayman, S.D. Incidence of *Salmonella* in poultry and meat products in Iraq. *Food Microbiol.* **1986**, *3*, 55–57. [CrossRef]

89. Noori, T.E.; Alwan, M.J. Isolation and identification of zoonotic bacteria from poultry meat. *Int. J. Adv. Res. Biol. Sci.* **2016**, *3*, 57–66.

90. Al-Zenki, S.; Al-Nasser, A.; Al-Safar, A.; Alomirah, H.; Al-Haddad, A.; Hendriksen, R.S.; Aarestrup, F.M. Prevalence and antibiotic resistance of *Salmonella* isolated from a poultry farm and processing plant environment in the state of Kuwait. *Foodborne Pathog. Dis.* **2007**, *4*, 367–373. [CrossRef]

91. Oueslati, W.; Rjeibi, M.R.; Mhadhbi, M.; Jbeli, M.; Zrelli, S.; Ettriqui, A. Prevalence, virulence and antibiotic susceptibility of *Salmonella* spp. strains, isolated from beef in Greater Tunis (Tunisia). *Meat Sci.* **2016**, *119*, 154–159. [CrossRef]

92. al-Nakhli, H.M.; al-Ogaily, Z.H.; Nassar, T.J. Representative *Salmonella* serovars isolated from poultry and poultry environments in Saudi Arabia. *Rev. Sci. Tech.* **1999**, *18*, 700–709. [CrossRef]

93. Ahmed, A.M.; Shimamoto, T. Isolation and molecular characterization of *Salmonella enterica*, *Escherichia coli* O157:H7 and *Shigella* spp. from meat and dairy products in Egypt. *Int. J. Food Microbiol.* **2014**, *168–169*, 57–62. [CrossRef]

94. Tarabees, R.; Elsayed, M.S.A.; Shawish, R.; Basiouni, S.; Shehata, A.A. Isolation and characterization of *Salmonella* Enteritidis and *Salmonella* Typhimurium from chicken meat in Egypt. *J. Infect. Dev. Ctries.* **2017**, *11*, 314–319. [CrossRef]

95. Amajoud, N.; Bouchrif, B.; El Maadoudi, M.; Skalli Senhaji, N.; Karraouan, B.; El Harsal, A.; El Abrini, J. Prevalence, serotype distribution, and antimicrobial resistance of *Salmonella* isolated from food products in Morocco. *J. Infect. Dev. Ctries.* **2017**, *11*, 136–142. [CrossRef]

96. Shah, A.H.; Korejo, N. Antimicrobial resistance profile of *Salmonella* Serovars isolated from chicken meat. *J. Vet. Anim. Sci.* **2012**, *2*, 40–46.

97. Gharieb, R.M.; Tartor, Y.H.; Khedr, M.H. Non-typhoidal *Salmonella* in poultry meat and diarrhoeic patients: Prevalence, antibiogram, virulotyping, molecular detection and sequencing of class I integrons in multidrug resistant strains. *Gut Pathog.* **2015**, *7*, 34. [CrossRef]

98. Soltan-Dallal, M.; Sharifi-Yazdi, M.K.; Mirzaei, N.; Kalantar, E. Prevalence of *Salmonella* spp. in packed and unpacked red meat and chicken in South of Tehran. *Jundishapur J. Microbiol.* **2014**, *7*, e9254. [CrossRef]

99. Malkawi, H. Molecular identification of *Salmonella* isolates from poultry and meat productsin Irbid City, Jordan. *World J. Microbiol. Biotechnol.* **2003**, *19*, 455–459. [CrossRef]

100. Nimri, L.; Abu Al-Dahab, F.; Batchoun, R. Foodborne bacterial pathogens recovered from contaminated shawarma meat in northern Jordan. *J. Infect. Dev. Ctries.* **2014**, *8*, 1407–1414. [CrossRef]

101. Jaradat, Z.W.; Abedel Hafiz, L.; Ababneh, M.M.; Ababneh, Q.O.; Al Mousa, W.; Al-Nabulsi, A.; Osaili, T.M.; Holley, R. Comparative analysis of virulence and resistance profiles of *Salmonella* Enteritidis isolates from poultry meat and foodborne outbreaks in northern Jordan. *Virulence* **2014**, *5*, 601–610. [CrossRef]

102. El Shrek, Y.M.; Ali, M.R. Microbiological study of spiced chicken burgers in Tripoli City, Libya. *East. Mediterr. Health J.* **2012**, *18*, 653–662. [CrossRef]

103. Soomro, A.H.; Khaskheli, M.; Bhutto, B.; Shah, G.; Azizullah, M.; Dewani, P. Prevalence and antimicrobial resistance of *Salmonella* serovars isolated from poultry meat in Hyderabad, Pakistan. *Turk. J. Vet. Anim. Sci.* **2010**, *34*, 455–460.

104. El Hussein, A.; Elmadiena, M.M.; Elsaid, S.; Siddig, M.; Muckle, C.A.; Cole, L.; Wilkie, E.; Mistry, K.; Perets, A. Prevalence of *Salmonella enterica* subspecies enterica serovars in Khartoum State, Sudan. *Res. J. Microbiol.* **2010**, *5*, 966–973. [CrossRef]

105. Hasan, S. Investigation of *Salmonella* in Retail Raw Meat in Syria. *J. Agric. Chem. Biotech.* **2013**, *4*, 147–153.

106. Abbassi-Ghozzi, I.; Jaouani, A.; Hammami, S.; Martinez-Urtaza, J.; Boudabous, A.; Gtari, M. Molecular analysis and antimicrobial resistance of *Salmonella* isolates recovered from raw meat marketed in the area of "Grand Tunis", Tunisia. *Pathol. Biol.* **2012**, *60*, e49–e54. [CrossRef] [PubMed]

107. Khan, M.A.; Suryanarayan, P.; Ahmed, M.M.; Vaswani, R.B.; Faheem, S.M. Antimicrobial susceptibility of *Salmonella* isolates from chicken meat samples in Dubai, United Arab Emirates. *Int. J. Food Nutr. Public Health* **2010**, *3*, 149–159.

108. Ozdemir, K.; Acar, S. Plasmid profile and pulsed-field gel electrophoresis analysis of *Salmonella enterica* isolates from humans in Turkey. *PLoS ONE* **2014**, *9*, e95976. [CrossRef] [PubMed]

109. Chang, Y.C.; Scaria, J.; Ibraham, M.; Doiphode, S.; Chang, Y.F.; Sultan, A.; Mohammed, H.O. Distribution and factors associated with *Salmonella enterica* genotypes in a diverse population of humans and animals in Qatar using multi-locus sequence typing (MLST). *J. Infect. Public Health* **2016**, *9*, 315–323. [CrossRef] [PubMed]

110. Dyson, Z.A.; Thanh, D.P.; Bodhidatta, L.; Mason, C.J.; Srijan, A.; Rabaa, M.A.; Vinh, P.V.; Thanh, T.H.; Thwaites, G.E.; Baker, S.; et al. Whole genome sequence analysis of *Salmonella* Typhi isolated in Thailand before and after the introduction of a national immunization program. *PLoS Negl. Trop. Dis.* **2017**, *11*, e0005274. [CrossRef]

111. Pornsukarom, S.; van Vliet, A.H.M.; Thakur, S. Whole genome sequencing analysis of multiple *Salmonella* serovars provides insights into phylogenetic relatedness, antimicrobial resistance, and virulence markers across humans, food animals and agriculture environmental sources. *BMC Genom.* **2018**, *19*, 801. [CrossRef] [PubMed]

112. Kwong, J.C.; Mercoulia, K.; Tomita, T.; Easton, M.; Li, H.Y.; Bulach, D.M.; Stinear, T.P.; Seemann, T.; Howden, B.P. Prospective Whole-genome sequencing enhances national surveillance of *Listeria monocytogenes*. *J. Clin. Microbiol.* **2016**, *54*, 333–342. [CrossRef] [PubMed]

113. Phillips, A.; Sotomayor, C.; Wang, Q.; Holmes, N.; Furlong, C.; Ward, K.; Howard, P.; Octavia, S.; Lan, R.; Sintchenko, V. Whole genome sequencing of *Salmonella* Typhimurium illuminates distinct outbreaks caused by an endemic multi-locus variable number tandem repeat analysis type in Australia, 2014. *BMC Microbiol.* **2016**, *16*, 211. [CrossRef] [PubMed]

114. Kang, M.-S.; Oh, J.-Y.; Kwon, Y.-K.; Lee, D.Y.; Jeong, O.M.; Choi, B.K.; Youn, S.Y.; Jeon, B.W.; Lee, H.J.; Lee, H.S. Public health significance of major genotypes of *Salmonella enterica* serovar Enteritidis present in both human and chicken isolates in Korea. *Res. Vet. Sci.* **2017**, *112*, 125–131. [CrossRef]

115. Cai, Y.; Tao, J.; Jiao, Y.; Fei, X.; Zhou, L.; Wang, Y.; Zheng, H.; Pan, Z.; Jiao, X. Phenotypic characteristics and genotypic correlation between *Salmonella* isolates from a slaughterhouse and retail markets in Yangzhou, China. *Int. J. Food Microbiol.* **2016**, *222*, 56–64. [CrossRef]

116. Thompson, C.K.; Wang, Q.; Bag, S.K.; Franklin, N.; Shadbolt, C.T.; Howard, P.; Fearnley, E.J.; Quinn, H.E.; Sintchenko, V.; Hope, K.G. Epidemiology and whole genome sequencing of an ongoing point-source *Salmonella* Agona outbreak associated with sushi consumption in western Sydney, Australia 2015. *Epidemiol. Infect.* **2017**, *145*, 2062–2071. [CrossRef]

117. Jagadeesan, B.; Gerner-Smidt, P.; Allard, M.W.; Leuillet, S.; Winkler, A.; Xiao, Y.; Leuillet, S.; Winkler, A.; Xiao, Y.; Chaffron, S.; et al. The use of next generation sequencing for improving food safety: Translation into practice. *Food Microbiol.* **2019**, *79*, 96–115. [CrossRef]

118. Hur, J.; Jawale, C.; Lee, J.H. Antimicrobial resistance of *Salmonella* isolated from food animals: A review. *Food Res. Int.* **2012**, *45*, 819–830. [CrossRef]

119. Mukerji, S.; O'Dea, M.; Barton, M.; Kirkwood, R.; Lee, T.; Abraham, S. Development and transmission of antimicrobial resistance among Gram-negative bacteria in animals and their public health impact. *Essays Biochem.* **2017**, *61*, 23–35. [CrossRef] [PubMed]

120. Yildirim, Y.; Gonulalan, Z.; Pamuk, S.; Ertas, N. Incidence and antibiotic resistance of *Salmonella* spp. on raw chicken carcasses. *Food Res. Int.* **2011**, *44*, 725–728. [CrossRef]

121. Ahmed, A.M.; Shimamoto, T.; Shimamoto, T. Characterization of integrons and resistance genes in multidrug-resistant *Salmonella enterica* isolated from meat and dairy products in Egypt. *Int. J. Food Microbiol.* **2014**, *189*, 39–44. [CrossRef]

122. Jassim, A.M. In-home drug storage and self-medication with antimicrobial drugs in Basrah, Iraq. *Oman Med. J.* **2010**, *25*, 79–87. [CrossRef] [PubMed]

123. Kambal, A.M. Antimicrobial susceptibility and serogroups of *Salmonella* isolates from Riyadh, Saudi Arabia. *Int. J. Antimicrob. Agents* **1996**, *7*, 265–269. [CrossRef]

124. Prakash, K.P. Epidemiology and antimicrobial resistance of enteric pathogens in Dhahira Region, Oman. *Iran. J. Public Health* **2008**, *37*, 60–69.

125. Jamal, W.Y.; Pal, T.; Rotimi, V.O.; Chugh, T.D. Serogroups and antimicrobial susceptibility of clinical isolates of *Salmonella* species from a teaching hospital in Kuwait. *J. Diarrhoeal Dis. Res.* **1998**, *16*, 180–186. [PubMed]

126. Chen, S.; Zhao, S.; White, D.G.; Schroeder, C.M.; Lu, R.; Yang, H.; McDermott, P.F.; Ayers, S.; Meng, J. Characterization of multiple-antimicrobial-resistant *Salmonella* serovars isolated from retail meats. *Appl. Environ. Microbiol.* **2004**, *70*, 1–7. [CrossRef] [PubMed]

127. Doublet, B.; Schwarz, S.; Kehrenberg, C.; Cloeckaert, A. Florfenicol resistance gene *floR* is part of a novel transposon. *Antimicrob. Agents Chemother.* **2005**, *49*, 2106–2108. [CrossRef]

128. Michael, G.B.; Schwarz, S. Antimicrobial resistance in zoonotic nontyphoidal *Salmonella*: An alarming trend? *Clin. Microbiol. Infect.* **2016**, *22*, 968–974. [CrossRef]

129. Zhu, Y.; Lai, H.; Zou, L.; Yin, S.; Wang, C.; Han, X.; Xia, X.; Hu, K.; He, L.; Zhou, K.; et al. Antimicrobial resistance and resistance genes in *Salmonella* strains isolated from broiler chickens along the slaughtering process in China. *Int. J. Food Microbiol.* **2017**, *259*, 43–51. [CrossRef]

130. Van, T.T.H.; Nguyen, H.N.K.; Smooker, P.M.; Coloe, P.J. The antibiotic resistance characteristics of non-typhoidal *Salmonella enterica* isolated from food-producing animals, retail meat and humans in South East Asia. *Int. J. Food Microbiol.* **2012**, *154*, 98–106. [CrossRef]

131. Campos, M.J.; Palomo, G.; Hormeño, L.; Herrera-León, S.; Domínguez, L.; Vadillo, S.; Píriz, S.; Quesada, A. Prevalence of quinolone resistance determinants in non-typhoidal *Salmonella* isolates from human origin in Extremadura, Spain. *Diagn. Microbiol. Infect. Dis.* **2014**, *79*, 64–69. [CrossRef]

132. Hopkins, K.L.; Wootton, L.; Day, M.R.; Threlfall, E.J. Plasmid-mediated quinolone resistance determinant *qnrS1* found in *Salmonella enterica* strains isolated in the, UK. *J. Antimicrob. Chemother.* **2007**, *59*, 1071–1075. [CrossRef]

133. Herrera-Leon, S.; Gonzalez-Sanz, R.; Herrera-Leon, L.; Echeita, M.A. Characterization of multidrug-resistant Enterobacteriaceae carrying plasmid-mediated quinolone resistance mechanisms in Spain. *J. Antimicrob. Chemother.* **2011**, *66*, 287–290. [CrossRef]

134. Zhao, S.; Blickenstaff, K.; Glenn, A.; Ayers, S.L.; Friedman, S.L.; Abbott, J.W.; McDermott, P.F. Beta-Lactam resistance in *Salmonella* strains isolated from retail meats in the United States by the National Antimicrobial Resistance Monitoring System between 2002 and 2006. *Appl. Environ. Microbiol.* **2009**, *75*, 7624–7630. [CrossRef] [PubMed]

135. Rotimi, V.O.; Jamal, W.; Pal, T.; Sonnevend, A.; Dimitrov, T.S.; Albert, M.J. Emergence of multidrug-resistant *Salmonella* spp. and isolates with reduced susceptibility to ciprofloxacin in Kuwait and the United Arab Emirates. *Diagn. Microbiol. Infect. Dis.* **2008**, *60*, 71–77. [CrossRef] [PubMed]

Article

Risk Factors for Zoonotic Tuberculosis at the Wildlife–Livestock–Human Interface in South Africa

Petronillah R. Sichewo [1,2,*], Anita L. Michel [1,3] , Jolly Musoke [4,5] and Eric M.C. Etter [6,7,8]

1   Department of Veterinary Tropical Diseases, Bovine Tuberculosis and Brucellosis Research Programme, Faculty of Veterinary Sciences, University of Pretoria, Private Bag X04, Onderstepoort 0110, Pretoria, South Africa
2   Department of Animal and Wildlife Sciences, Faculty of Natural Resources Management and Agriculture, Midlands State University, P. Bag 9055, Gweru, Midlands 00263, Zimbabwe
3   Research Associate at the National Zoological Gardens of South Africa, Pretoria 0001, South Africa
4   National Health Laboratory Services, Department of Medical Microbiology, Universitas, Bloemfontein 9301, South Africa
5   Department of Medical Microbiology, Faculty of Health Science, University of the Free State, Bloemfontein 9301, South Africa
6   Department of Production Animal Studies, Faculty of Veterinary Sciences, University of Pretoria, Private Bag X04, Onderstepoort 0110, South Africa
7   CIRAD, UMR Animal, Santé, Territoires, Risque et Ecosystèmes (ASTRE), 34398 Montpellier, France
8   ASTRE, Univ Montpellier, CIRAD, INRA, 34398 Montpellier, France
*   Correspondence: psichewo@gmail.com; Tel.: +27-774-356-3943

Received: 29 May 2019; Accepted: 21 June 2019; Published: 14 July 2019

Abstract: A cross-sectional study was conducted to investigate the risk factors associated with zoonotic tuberculosis in humans and its transmission to people living at the wildlife–livestock–human interface. A questionnaire was administered to collect information on food consumption habits, food handling practices, and knowledge of zoonotic TB. Sputum samples were also collected from 150 individuals that belonged to households of cattle farmers with or without a bTB infected herd. In addition, 30 milk samples and 99 nasal swabs were randomly collected from cattle in bTB infected herds for isolation of *Mycobacterium bovis* (*M. bovis*). The sputum samples were screened for TB using the GeneXpert test and this was followed by mycobacterial culture and speciation using molecular techniques. No *M. bovis* was isolated from TB positive sputum samples and only one sample was confirmed as *Mycobacterium tuberculosis* (*M. tuberculosis*). *M. bovis* was isolated from 6.6% (n = 2/30) milk samples and 9% (n = 9/99) of nasal swabs. Ownership of a bTB infected herd and consumption of milk were recognized as highly significant risk factors associated with a history of TB in the household using multiple correspondence analysis (MCA) and logistic regression. The findings from this study have confirmed the potential for zoonotic TB transmission via both unpasteurized milk and aerosol thus, the role of *M. bovis* in human TB remains a concern for vulnerable communities.

Keywords: bovine tuberculosis (bTB); multiple correspondence analysis (MCA); *Mycobacterium bovis* (*M. bovis*); risk factors; wildlife–livestock–human interface; zoonotic TB

---

1. Introduction

Cattle are the maintenance host of Mycobacterium bovis (*M. bovis*) but many other domestic and wildlife animals can be affected by the pathogen [1]. *M. bovis* belongs to the Mycobacterium tuberculosis complex (MTC), a group of closely related organisms that causes TB in mammals including humans [2]. Although *M. tuberculosis* is the most common cause of human TB cases, an unspecified fraction occurs due to *M. bovis* infection and is referred to as zoonotic TB [3–5].

The true impact of *M. bovis* on the human TB epidemic is unclear due to the lack of routine bovine TB surveillance data from humans and resources for the identification of *M. bovis* [6,7]. In 2016 it was estimated that 147,000 new cases of zoonotic TB were reported in people, and 12,500 deaths due to the disease globally and most of these occurred in Africa [8]. Using the data that was available at the time of the study, Müller et al. (2013) estimated that *M. bovis* might be responsible for up to 37.7% of all human TB cases in Africa [6]. In studies carried out in several African countries *M. bovis* has been isolated from lymph node biopsies or aspirates of TB patients, in Uganda at a prevalence of 7%, in Tanzania 16% and in Ethiopia 17% [9–11]. While recent investigations in Zambia and Uganda diagnosed *M. bovis* at a prevalence of < 1% and < 3%, respectively from sputum samples of pulmonary TB patients [12,13].

Zoonotic TB is primarily acquired through the consumption of unpasteurized milk and dairy products, less frequently from eating of raw or improperly cooked infected meat and via aerosols inhaled from infected animals during direct human–livestock contact [5,14]. This supports the observation that zoonotic TB occurs more frequently as extra-pulmonary TB (EPTB) (9.4%) in humans than in the pulmonary form (2.1%) as mentioned by Etter et. al., (2006) [15]. The 'test and slaughter policy', compulsory pasteurization of milk and abattoir surveillance have been successfully implemented in developed countries leading to the near elimination of the disease in cattle and human populations [6]. Unlike in developing countries, particularly in Africa, these policies are absent or inadequate due to lack of resources, thus bTB is widespread in animals [16].

Risk factors for *M. bovis* transmission to people include demographic factors (e.g., number of family members, age), feeding habits, people living in close contact with their animals, socio-economic status, illiteracy (lack of knowledge of zoonotic TB), customs, traditions coupled with the increasing prevalence of HIV/AIDS pandemic [5,14,16,17]. Nevertheless, the principal risk factors that govern bTB epidemiology and transmission to humans living at the wildlife–livestock–human interface remain largely unknown in sub-Saharan Africa including South Africa.

In northern KwaZulu-Natal, in addition of bTB being endemic in wildlife in the game reserves such as the Hluhluwe iMfolozi, the disease was previously detected in cattle in the surrounding communal area with an overall herd prevalence of 28% [18,19]. The community is in this province with one of the highest TB incidence rate of 685/100,000) and an HIV prevalence of 27% in the country, where consumption of unboiled milk is a common practice, indicative of a significant role of *M. bovis* in human TB [20–22]. In addition, there is a presumptive lack of knowledge among livestock farmers on zoonotic TB and protective practices against zoonotic TB transmission have never been examined in the area. Against this background, there was a need to investigate the possible risk factors for bovine tuberculosis transmission from cattle to humans in a One Health approach and to assess the farmers' knowledge of zoonotic TB. The findings from this study will assist the policy makers in the selection of appropriate preventive animal health measures and designing of awareness programs led by community health workers.

2. Results

2.1. Household Study

Sputum and serum samples were collected from 75 individuals from households that owned bTB test positive cattle, similarly 75 samples were collected from households that owned bTB test negative cattle. This was less than the expected sample size of 300 and we attribute the low numbers to the absence of some family members during the times we visited their homesteads, and the unwillingness of either individuals or guardians (for children below the age of 15) to participate. Most of the individuals that participated were mostly men 62% (94/150) aged between 16–64 years as shown in Table 1. According to the cultural practices in the area, it is mostly the men that are involved in livestock activities.

Table 1. *Mycobacterium tuberculosis* complex and HIV status of household members based on bTB herd status.

bTB Herd Status	Age Group	No. of Participants	TB Positive (GeneXpert)	Confirmed *M.tuberculosis*	HIV Prevalence
bTB positive herd	16–64 years	68	5	1	40% (30/75)
	>64 years	7	2	-	-
bTB negative herd	16–64 years	72	3	-	33% (25/75)
	>64 years	3	-	-	-

M. bovis was not isolated from individuals in the study population. Out of the 10 samples that were TB positive from the GeneXpert test none of the samples were identified as *M. bovis*; only one sample was confirmed as *M. tuberculosis* through culture and deletion analysis of the RD4 and the rest of the samples were negative for mycobacterial organisms as shown in Table 1. The estimated population prevalence of *M. bovis* was 1.33% confidence interval (C.I) [0;3.92] and < 3.92% respectively for participants from households that owned bTB positive and uninfected herds. The GeneXpert test is the standard TB diagnostic in South Africa as recommended by the World Health Organisation for diagnosis of pulmonary TB [23]. The Gene Xpert (Cepheid), identifies *Mycobacterium tuberculosis* complex organism and determines drug resistance to rifampicin in TB positive samples. The overall HIV prevalence was 36% (individuals aged between 16–64) and there was no significant difference in HIV infection between the farmers that owned bTB infected and uninfected herds (Chi-squared test: *p* value = 0.5). Results in Table 1 summarizes the TB and HIV results from the household study. The participants that were unaware of their TB status (6%) or HIV status (10%) prior to enrolling in this study were informed of their results by the Department of Health at their respective clinics and received appropriate treatment.

2.2. Bovine Samples

A total of 30 milk samples and 99 nasal swabs were collected from bTB infected herds. The milk yield was low, and some lactating animals had ceased milk production due to the drought in the study area at the time of sampling. *M. bovis* was isolated from 9% (n = 9) of nasal swabs from infected herds and from 6.6% (n = 2) of the collected milk samples.

2.3. Questionnaire Survey

A total of 71 participants took part in the questionnaire survey, comprising 59% (n = 42) from households with bTB infected herds and 41% (n = 29) from households that owned uninfected herds. This is less than the calculated number of farmers (100) and this was due to the unavailability of family members because of work commitments and livestock herding activities. In a few cases there were elderly or young people at home who could not answer questions logically and people not willing to participate in the study.

2.4. Descriptive Analysis

Of the 71 participants, 40% (n = 31) reported a history of TB diagnosis in the family and 60% (n = 40) had no history of TB diagnosis in the family. The household demographics are summarized in Table 2 below.

Most of the participants were involved in the herding (100%) and milking of cows (86%). The participants reported consumption of milk (98%), especially as sour milk (89%) and this involved all the family members in 97% of the households. Most of the households obtained meat from their own cattle (94%) but none involved veterinary services for the inspection of the meat. More than 50% of the households had observed abnormal spots on organs that looked like TB lesions (as shown on the pictures by the interviewer) at least once during informal slaughtering. The family members of cattle owning households displayed poor knowledge of the disease transmission modes with 56%

being unaware of the zoonotic aspect of bovine tuberculosis and 63% being unable to give examples of protective practices during slaughter of animals. This is summarized in Table 3.

Table 2. Summary of household demographics determined by questionnaire survey.

Variable	Level	Response (n = 71) %
Gender	Male	66
	Female	34
Age group	16–64 years	87
	>64 years	13
Status in the family	Cattle owner	31
	Cattle keeper	11
	Member of the household	58
Education	No Education	18
	Primary education	23
	Secondary education	17
	High school	34
	Tertiary	8

Table 3. Risk factors and awareness of bTB as determined by questionnaire.

Category	Variable	Level	Responses (n = 71) %
Food consumption habits	Do you regularly consume milk from your animals?	Yes	98
		No	2
	Who mainly consumes milk in the household?	Whole household	97
		<12	1.5
		>64 years	1.5
	How often is milk consumed?	Daily	3
		Weekly	63
		When needed	34
	How do you consume your milk?	Soured (Amasi)	89
		Raw	6
		Boiled	4
	What is your source of meat?	Supermarket	58
		Own cattle	94
		Buy from others	17
	How do you process meat with abnormal spots?	Never seen spots	48
		Throw away meat	25
		Overcook meat	13
		Normal use of meat	11
	What is your source of water?	Boiling	3
		Own well	27
		Borehole	59
		Communal	14

2.5. Risk Factor Analysis

The analysis of risk factors was done using the history of TB in the family as reference variable (MCA) or the fixed (outcome) variable (Fisher test and general linear model).

2.5.1. Univariate Analysis and Logistic Regression of Risk Factors for bTB Transmission to Humans

Out of the 20 variables that were analysed using univariate analysis, five significant factors were identified (p value < 0.25) namely: consumption of milk weekly and when needed, consumption of meat from own cattle or from other farmers and ownership of a bTB positive herd as shown in Table 4.

Table 4. Univariate analysis of bTB transmission to people in cattle owning households.

Factor	*p* Value	Odds Ratio (95% CI)
Weekly consumption of milk	0.31	0.24 (0.04–3.22)
Consumption of milk when needed	0.20	2.14 (0.70–6.74)
Purchase of meat from other farmers	0.21	0.37 (0.06–1.70)
bTB positive herd	<0.01	10.8 (2.97–51.08)

2.5.2. Multiple Correspondence Analysis

The risk factors/practices that were closely related to the history of TB diagnosis in a family are: ownership of a bTB positive herd, not purchasing meat from other farmers, ownership of herds that belonged to Mpempe dip tank and farmers that had never seen abnormalities or TB like lesions on cattle organs during slaughter. Absence of history of TB diagnosis in the family was closely related to the ownership of an uninfected herd and previous observation of abnormalities or TB like lesions on organs from their animals. This is illustrated in Figure 1 depicting the history of TB diagnosis (positive) and related risk factors/practices inside the red circle and absence of history of TB diagnosis (negative) in the family and related protective factors/practices inside the blue circle.

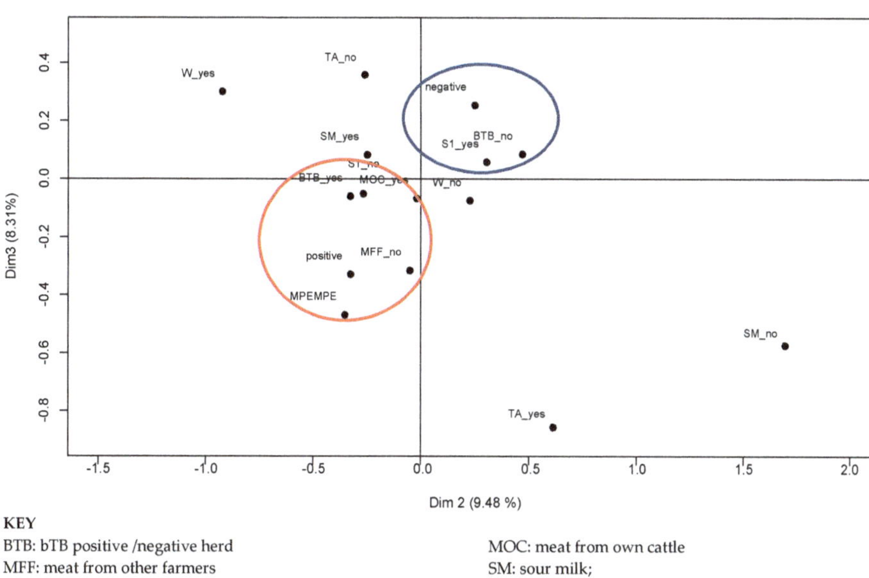

KEY

BTB: bTB positive /negative herd
MFF: meat from other farmers
S1: spots/TB like lesions
W: weekly consumption of milk
Mpempe: Mpempe dip tank

MOC: meat from own cattle
SM: sour milk;
TA: awareness of TB transmission
Positive: history of TB in the family
Negative: no history of TB in the family

Figure 1. Multiple correspondence analysis of risk factors associated with history of TB diagnosis in the family (positive or negative).

2.5.3. Logistic Regression of Risk Factors for bTB Transmission to Humans

All the statistically significant factors from the univariate analysis (*p* < 0.3) and other risk factors of biologically importance were included in the logistic regression model; the significant risk factors identified were, ownership of a bTB positive herd and consumption of milk from own cattle while consumption of meat from own cattle or other farmers were identified as protective practices as presented in Table 5.

Table 5. Significant risk factors for bTB transmission to people that live in the households that own cattle from logistic regression.

Factor	p Value	Adjusted OR	CI$_{95\%}$
Bovine TB positive cattle	0.00009 ***	29.308	6.495–208.361
Consumption of milk when needed	0.0393 *	4.937	1.197–27.130
Consumption of meat from own cattle	0.0493 *	0.053	0.001–0.755
Consumption of beef bought from other farmers	0.0674	0.105	0.005–0.866
Awareness of bTB transmission	0.0471 *	4.419	1.101–22.067

Notes: * $p < 0.05$ *** $p < 0.001$.

3. Discussion

The aim of this study was to investigate *M. bovis* infection in humans exposed to bTB infected cattle and the risk factors associated with zoonotic TB in a community living at the wildlife–livestock–human interface. *M. bovis* excretion was confirmed in milk (6.6%) and nasal discharges (9%) and the associated risk factors identified included consumption of unpasteurized milk particularly as sour milk and ownership of bTB infected cattle. The actual excretion rate in milk and nasal discharges should, however, be deemed higher due to loss of viable organisms during sample processing, the intermittent shedding of the pathogen by infected animals and the sensitivity of the culture method [24,25].

Despite the community's apparent exposure risk, *M. bovis* was not detected in sputum samples collected from individuals that belonged to households that owned cattle. This raises the question whether *M. bovis* is not transmitted in high enough numbers to infect humans or whether the loss of live organisms was too severe and reduced the viable bacterial concentration below the limit of detection, or alternatively whether the threshold of infection in cattle was too low to cause disease in humans. The latter is because bTB prevalence in the cattle populations under study was reported as 12% while the bTB prevalence in cattle in Britain, which was 40% in the 1940s, was linked to a 6% *M. bovis* infection burden in human TB [18,26,27]. Similar results have also been reported in Ethiopia at a wildlife–livestock–human interface with low bTB prevalence rate in cattle [28]. Another possible explanation for the absence of *M. bovis* infection in humans might be due to life-long protection against *M. bovis* as induced by Bacillus Calmette-Guerin (BCG) vaccination at birth. However, further evaluations have revealed that BCG's protective efficacy against *M. tuberculosis* infection and progression to disease is not absolute [29]. It has also been proposed that humans are more resistant to *M. bovis* as compared to *M. tuberculosis* as explained by reduced transmissibility and a lower risk of human disease establishment after infection [27,30].

Considering consumption of milk was identified as a significant risk factor and the apparent excretion of *M. bovis* in milk by lactating animals, non-pulmonary TB is highly likely to affect the exposed individuals. Therefore, sputum might have been an unsuitable sample for the detection of *M. bovis* in people that might be affected by non-pulmonary forms of TB as discussed by Debi et al. 2014 [31]. On the other hand, the participants had no apparent TB clinical signs such as a productive chronic cough, thus they experienced it challenging to produce sputum and instead inappropriate samples of saliva were collected which could have reduced the chance of detecting an underlying *M. bovis* infection. This was also supported by the discrepancy in the GeneXpert and culture results whereby, 10 sputum samples were GeneXpert positive and yet only one sample was positive for *M. tuberculosis* using culture. The GeneXpert test is a PCR based test that depends on the pathogen's DNA in saliva hence displays a higher sensitivity and specificity than culture [32], whereas, culture results are affected by the number and viability of bacteria and quality of the sputum processed [33] which was influenced by the distance and transport delays between collection and processing of sputum as the study was in a remote area.

In corroboration with our results, a study carried out in the Serengeti ecosystem of Tanzania did not find *M. bovis* from the sputa of TB patients despite the presence of the infection in cattle and wildlife [34]. Similarly, studies in Brazil, Uganda and Cote d'Ivoire found no *M. bovis* from human samples that included sputa despite bTB being prevalent in the cattle populations of these countries [12,35,36]. In contrast, *M. bovis* has been detected in sputum samples of livestock traders in Nigeria (10%) and pastoral communities in Ethiopia (15%) from both sputum and fine needle aspirates specimens [37,38]. The high prevalence in Ethiopia could be explained by cattle farmers sharing shelter with their livestock while in this study, none of the farmers shared living space with animals [17,38]. Findings from these individual studies indicate that zoonotic TB is a disease of public health importance particularly in poor-resourced communities that should not be ignored [6].

In this study, consumption of milk and meat from own cattle were identified as highly significant risk factors for TB in people. Moreover, most of the households (89%) reported consumption of milk as sour milk by almost everyone in the households (97%), predisposing the individuals to *M. bovis* infection. Consumption of soured milk frequently (weekly), as indicated by two thirds of the respondents may increase the risk of repeated *M. bovis* exposure for a higher number of consumers. It has been demonstrated that *M. bovis* persists for up to 14 days in soured milk depending on the initial bacterial concentration in raw milk, souring and storage temperature. This is further exacerbated by the commonly practiced pooling and supplementation of milk with left over sour milk "stock" for a continuous production of sour milk that results in the contamination of milk from uninfected animals [39]. The results suggest ingestion of contaminated food as the most important route of infection and this would primarily result in extra-pulmonary TB, making it highly unlikely to detect *M. bovis* infection in sputa. *M. bovis* infection is frequently associated with extra-pulmonary as demonstrated by studies in Tanzania and Ethiopia where *M. bovis* was more prevalent in cases of extra-pulmonary TB than pulmonary TB [10,38]. However, in the present study cases of lymphadenitis (EPTB) were not encountered. The initial *M. bovis* load from the lactating cows might be too low to survive souring hence absence of EPTB in people as supported by the absence of the bacteria from the culture of milk samples that were collected sporadically from the infected cows in the study area over a period of 12 months [40].

All the farmers reported absence of meat inspection by veterinary public health officers during slaughter and 48% had no knowledge of the TB lesions that characterised organs of an infected animal. Inspection of meat at slaughter is of significance to control the spread of *M. bovis* infection from infected cattle to humans through removal of contaminated organs or condemnation of carcasses with disseminated bTB. It also facilitates the trace-back of *M. bovis* infection to herd level in eradication programs such as the one that was implemented in European countries in the 1960s to control zoonotic TB in humans [27]. Consumption of meat from the supermarket that undergoes regular inspection at the abattoirs was reported by more than half of the families (58%). In addition, consumption of meat from own cattle or bought from other farmers were indicated as protective practices. The consumption of undercooked contaminated meat has been previously reported in other investigations as a potential risk factor for transmission of Mycobacteria and other zoonotic diseases to humans although the public health significance is yet to be quantified [5,14,41–43].

The households that owned uninfected herds were associated with the absence of previous diagnosis of human TB but also with reporting of previous observations of organ abnormalities in slaughter animals. Equipped with this knowledge these households may be less likely to consume infected meat and thus be better protected against food borne zoonoses. These findings support the assumption that knowledge of bTB in cattle and its control is linked to a reduced risk to zoonotic TB.

Previous studies have reported ownership of cattle, direct contact with animals and living in close proximity with animals as important drivers of zoonotic TB transmission to people [43–45]. In this study, livestock keeping activities were not identified as risk factors although all the respondents participated in at least one livestock keeping activity such as milking, herding or examination of animals, none of them slept with or near their animals. We conclude that transmission of *M. bovis* through direct contact was less likely to occur in our study area due to the short periods of exposure of individuals in a confined environment to infected herds. In contrast, the livestock traders of Nigeria trade in a congested environment for long periods of time with increased human-to-cattle contact leading to zoonotic TB [37].

Approximately two thirds of the respondents were not aware of zoonotic TB transmission from cattle to humans. The lack of understanding of disease transmission precludes protective practices that contribute to effective disease control programs. In contrast with other studies elsewhere which demonstrated a positive correlation of zoonotic TB knowledge with post primary education this was apparently not the case in our study where 59% of the participants had post primary education [43,46].

4. Materials and Methods

4.1. Study Area

The study was conducted from August to September 2017, in four villages from Big 5 False Bay Municipality in uMkhanyakude district, northern Kwa-Zulu Natal province, South Africa. The municipality is situated between game reserves that include iSimangaliso Wetland Park (formerly St. Lucia) and Hluhluwe iMfolozi Park (HiP). Most of the land in the municipality is used for subsistence farming, game lodge activities and the north-eastern parts of the municipality are occupied by densely settled rural traditional communities. The study involved the cattle owners from four villages (Mnqobokasi, Makhasa, Mduku and Nibela), that were associated with the four dip tanks respectively (Masakeni, Mpempe, Nkomo, Nibela) where bTB testing of cattle had been carried out as part of a research project in 2016 and 2017 [18].

4.2. Study Population

The study population was recruited from the households that owned cattle herds that belonged to one of the 4 dip tanks. Participation in the study was voluntary, participants were approached through the community health care givers and the purpose of the study was explained. The case-control study involved two groups of households; 50 households that owned bTB infected herds and an equal number of cattle farmers that owned uninfected herds. Three family members were recruited from each household representing the three age categories: adolescents below 15 years, adults 16–64 years and the elderly above 64 years. The total number of participants calculated was 300, comprising of individuals that included head of households (cattle owners), herd boys/man/women (cattle keepers) and female family members of the cattle owning households.

4.3. Household Study

At household level samples (sputum and serum) were collected from three members of the family and a questionnaire was administered to one member of the household, either the cattle owner or cattle keeper or female household member.

4.3.1. Collection and Processing of Human Samples

The research team visited each of the selected households and collected two sputum samples of approximately 2–5 mL into sterile, leak proof plastic container and 3–6 mL of blood into serum separating tubes (SST) from each participant. The name, age and gender for each participant was recorded but for the purpose of the study and confidentiality a unique code was allocated to each participant and served as an identifier on the sample containers. The samples were transported to the local district hospital (Mseleni NHLS laboratory) in a cold chain at 4 °C for the GeneXpert test.

The GeneXpert test is the standard TB diagnostic in South Africa as recommended by the World Health Organisation for diagnosis of pulmonary TB [23]. The Gene Xpert (Cepheid), identifies *Mycobacterium tuberculosis* complex organism (does not differentiate *M. bovis* and *M. tuberculosis*) and determines drug resistance to rifampicin in TB positive samples. The 2nd sputum samples from all the TB positive samples were transported in equal volumes of cetylpyridinium chloride (CPC) (1%) in a cold chain for mycobacterial culture and *M. bovis* identification at the National

Health Laboratory Services (NHLS) Medical Microbiology University of the Free State. The serum samples were transported to Inkosi Albert Luthuli hospital-NHLS laboratory for HIV testing using the Enzyme-linked immunosorbent assay (ELISA) serological test. The sample collection and processing procedure is shown in Figure 2.

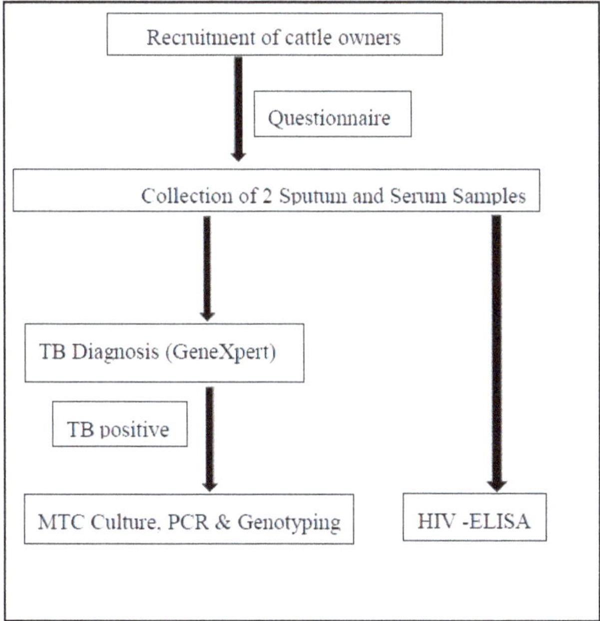

Figure 2. A flow chart of sample collection from households of cattle farmers (including cattle owners and cattle keepers) and laboratory processing for human samples (sputum and serum) and questionnaire administration.

4.3.2. Mycobacterium Species Culture and Identification

Sputum samples from participants found to be positive for TB during the screening test were inoculated into Mycobacterial growth indicator tube (MGIT) broth, incubated at 37 °C in a BD Bactec MGIT 320 with daily monitoring for 28 days and then once a week for up to 42 days. MGIT broth supports the growth of both *M. bovis* and *M. tuberculosis*. DNA was extracted from growth positive cultures using the Hain Genolyse kit reagents according to the manufacturer's instructions. Briefly, 1 mL of the MGIT culture was centrifuged at 1000 g for 15 min, the pellet resuspended in 100 uL of lysis buffer followed by incubation for 5 min at 95 °C. For neutralisation, 100 uL of the buffer was added, centrifuged for 5 min at maximum speed and the supernatant stored as DNA for subsequent reactions. The DNA was used as a template in deletion analysis using PCR reaction for confirmation of mycobacterial growth as *M. bovis* targeting the region of difference (RD), RD4 as previously outlined by Warren et al. (2006) [47].

4.3.3. Questionnaire Survey

A pre-tested questionnaire with open-ended and closed questions was administered for 20–30 min (face-to-face interviews) to one household member above 18 years that was involved with livestock keeping activities. The questionnaire was developed in English and was translated into the local language of isi-Zulu. The questionnaire was divided into four sections. The first section detailed demographic characteristics, and this included age, sex, level of education and number of family members. The second section prompted the interviewee to respond to questions pertaining to livestock management activities such as frequency of human-to-cattle contact during activities such as herding, milking and examination of the animals. The third section invited the respondents to answer questions on their food consumption behavior, food handling and processing practices. The questions in the fourth section were on the knowledge and awareness of bovine tuberculosis in cattle, humans and the history of TB in the family for the past 12 months.

4.4. Collection of Cattle Samples

Nasal swabs and milk samples were collected from cattle from bTB infected herds as previously determined by the interferon gamma assay (IFN-γ) (BOVIGAM®) [18]. Households with bTB positive herds brought their animals to the dip tank, milk was systematically collected from lactating animals and nasal swabs through random sampling of the animals. From a herd with >10 animals, every 5th animal was sampled while for herds < 10 every 3rd animal was sampled.

4.4.1. Collection and Processing of Milk Samples

Verbal consent was obtained from the cattle owners before milking commenced. We intended to collect 50 mL of milk per animal, but the actual quantities collected depended on the availability of milk from the lactating animals. The milking was done manually, and 20–40 mL of milk was collected into sterile screw-capped 50 mL sterile centrifuge tubes by the cattle owner or cattle keeper after rinsing their hands and udder with 70% alcohol. The tubes were labelled with a unique code that identified the herd and placed in a cool box for transportation to the local state veterinary laboratories where it was stored at −20 °C until processing at University of Pretoria-Department of Veterinary Tropical Diseases BSL 2+ laboratory.

Milk decontamination was done using cetylpyridinium chloride (CPC) as previously described by Michel et al. [39]. The sediments were inoculated onto Löwenstein Jensen slopes supplemented with pyruvate and antibiotics cocktail (polymyxin B (200 IU/ml), amphotericin B (10 μg/mL) carbenicillin (100 μg/mL) and trimethoprim (10 μg/mL)) and incubated at 37 °C for 10 weeks with weekly monitoring for bacterial growth.

4.4.2. Collection and Processing of Nasal Swabs

Nasal swabs were collected from bTB infected cattle using a 50-cm long homemade sterile swab made of aluminum wire and gauze from one nasal passage per animal. These were immediately immersed and expunged into 25 mL of sterile phosphate buffer saline (PBS) at pH = 7 in 50 mL centrifuge tubes in the field. The samples were kept at 4 °C until they were processed for *M. bovis* isolation and identification. The decontamination of nasal swabs was performed using 2% HCl as described by Gcebe et al. [48]. The sediments obtained after decontamination were incubated for 20 hours after adding 1 mL of 50 μg/mL of amphotericin B (antifungal). The solution was then inoculated onto Löwenstein Jensen media with pyruvate and a cocktail of antibiotics (polymyxin B (200 IU/ml), amphotericin B (10 μg/mL) carbenicillin (100 μg/mL) and trimethoprim (10 μg/mL) (NHLS, South Africa)) and incubated at 37 °C for 10 weeks.

4.4.3. Mycobacterium Bovis Identification

Crude DNA extraction from *M. bovis* isolates was done by boiling a loopful of cells in 100 μL of distilled water using a heating block for 25 min at 95 °C [49]. *M. bovis* was confirmed in the isolates using deletion analysis as described by Warren et al., targeting RD4 and RD9 primers recognized by the amplification of DNA products with band sizes 268 bp and 108 bp, respectively [47].

4.5. Statistical Analysis

All the data collected during the household study, questionnaire survey, bovine and milk samples results were stored in Microsoft Excel. The data from the questionnaire survey were exported and analysed with the R software (© 3.4.4, 2018, R Foundation for Statistical Computing, Vienna, Austria). Bovine TB transmission from cattle to humans was initially examined using general descriptive analysis. The effect of potential risk factors on the variable history of TB diagnosis in members of cattle owning households was analysed using the univariate analysis by means of the two-tailed Fischer's exact test. The association of explanatory variables with a history of TB patients in the households was further described using the multiple correspondence analysis (MCA) [50]. The effect of potential risk factors on TB in people was analysed using univariate analysis using the two-tailed Fischer's exact test. Logistic regression analysis was performed to quantify these risk factors (Generalised Linear model (GLM-family = binomial) using the significant risk factors from the univariate analysis ($p < 0.25$).

4.6. Ethical Clearance

Animal ethical and biosafety transport clearances were obtained from the University of Pretoria Animal Ethics committee (Ref: V078-16) and the Department of Agriculture Forestry and Fisheries under Section 20 (12/11/1/1/6/1). Ethical clearance for human study was obtained from the University of Pretoria Ethics committees in the Faculty of Health Sciences (321/2016) for sputum and sera collection and Faculty of Humanities (GWO170814HS) for the questionnaire survey. The study was also explained to the participants before collection of samples and verbal and written consent obtained.

5. Conclusions

The isolation of *M. bovis* from milk and nasal secretions confirms that there is a potential risk of bovine tuberculosis transmission from cattle to humans through exposure to respiratory exudates, from aerosols and supported by the consumption of contaminated raw and sour milk. Nevertheless, the current study could not confirm *M. bovis* transmission between animals and humans at the wildlife–livestock–human interface. Therefore, in this population there was no link between food consumption practices and transmission of zoonotic TB to people. However, the presence of *M. bovis* in animal products that are known to be consumed by farmers without specific precaution is of significance in the designing of control and management measures that will reduce the impact of the disease on human health.

Author Contributions: Conceptualization, A.L.M.; E.M.C.E.; Resources supervision, A.L.M.; J.M., P.R.S.; E.M.C.E.; Methodology, A.L.M. and E.M.C.E.; Formal analysis, P.R.S. and E.M.C.E.; Investigation, P.R.S. and J.M.; Writing—original draft preparation, P.R.S.; Writing—review and editing, P.R.S.; J.M.; A.L.M.; E.M.C.E.; Funding acquisition, A.L.M. and J.M.

Funding: This research was funded by The National Research Foundation grant number 113594, The National Research Foundation, Thuthuka grant ref TTK160525166191 and the Institute of Tropical Medicine, Belgium for the funding received through the grant agreement with the Department of Veterinary Tropical Diseases, University of Pretoria.

Acknowledgments: We thank the health authorities from Department of Health-province of Kwa-Zulu Natal, uMkhanyakude district and Mseleni Hospital. The laboratory staff, nurses from the local clinics and community care givers are acknowledged for their kind assistance. We are grateful to the farmers of Big 5 False Bay Municipality, uMkhanyakude district for consenting to participate in our study.

Conflicts of Interest: The authors declare no conflict of interest.

References

1. Michel, A.L.; Müller, B.; Van Helden, P.D. *Mycobacterium bovis* at the animal–human interface: A problem, or not? *Vet. Microbiol.* **2010**, *140*, 371–381. [CrossRef] [PubMed]
2. Galagan, J.E. Genomic insights into tuberculosis. *Nat. Rev. Genet.* **2014**, *15*, 307–320. [CrossRef] [PubMed]
3. WHO. *Global Tuberculosis Report*; WHO: Geneva, Switzerland, 2018.

4.   Miller, M. *Tuberculosis in South African Wildlife: Why Is It Important*; Sun Media: Stellenbosch, South Africa, 2015.

5.   Gumi, B.; Schelling, E.; Berg, S.; Firdessa, R.; Erenso, G.; Mekonnen, W.; Hailu, E.; Melese, E.; Hussein, J.; Aseffa, A.; et al. Zoonotic Transmission of Tuberculosis Between Pastoralists and Their Livestock in South-East Ethiopia. *EcoHealth* **2012**, *9*, 139–149. [CrossRef] [PubMed]

6.   Müller, B.; Dürr, S.; Alonso, S.; Hattendorf, J.; Zinsstag, J.; Laisse, C.J.; Parsons, S.D.; Van Helden, P.D. Zoonotic *Mycobacterium bovis*-induced tuberculosis in humans. *Emerg. Infect. Dis.* **2013**, *19*, 899–908. [CrossRef] [PubMed]

7.   Olea-Popelka, F.; Muwonge, A.; Perera, A.; Dean, A.S.; Mumford, E.; Erlacher-Vindel, E.; Raviglione, M. Zoonotic tuberculosis in human beings caused by *Mycobacterium bovis*—A call for action. *Lancet Infect. Dis.* **2017**, *17*, e21–e25. [CrossRef]

8.   World Health Organization (WHO). Food and Agriculture Organisation of the United Nations (FAO); World Organisation for Animal Health (OIE) IUAT and International Union Against Tuberculosis and Lung Disease (The Union). Road map for zoonotic tuberculosis. In Proceedings of the 48th Union World Conference on Lung Health, Guadalajara, Mexico, 11–14 October 2017.

9.   Kidane, D.; Olobo, J.O.; Habte, A.; Negesse, Y.; Aseffa, A.; Abate, G.; Yassin, M.A.; Bereda, K.; Harboe, M. Identification of the Causative Organism of Tuberculous Lymphadenitis in Ethiopia by PCR. *J. Clin. Microbiol.* **2002**, *40*, 4230–4234. [CrossRef] [PubMed]

10.   Kazwala, R.R.; Daborn, C.J.; Sharp, J.M.; Kambarage, D.M.; Jiwa, S.F.; Mbembati, N.A. Isolation of *Mycobacterium bovis* from human cases of cervical adenitis in Tanzania: A cause for concern? *Int. J. Tuberc. Lung Dis.* **2001**, *5*, 87–91. [PubMed]

11.   Oloya, J.; Opuda-Asibo, J.; Kazwala, R.; Demelash, A.B.; Skjerve, E.; Lund, A.; Djonne, B. Mycobacteria causing human cervical lymphadenitis in pastoral communities in the Karamoja region of Uganda. *Epidemiol. Infect.* **2008**, *136*, 636–643. [CrossRef] [PubMed]

12.   Byarugaba, F.; Etter, E.M.C.; Godreuil, S.; Grimaud, P. Grimaud, P. Pulmonary tuberculosis and *Mycobacterium bovis*, Uganda. *Emerg. Infect. Dis.* **2009**, *15*, 124–125. [CrossRef]

13.   Malama, S.; Muma, J.B.; Olea-Popelka, F.; Mbulo, G. Isolation of *Mycobacterium bovis* from Human Sputum in Zambia: Public Infectious Diseases & Therapy Isolation of *Mycobacterium bovis* from Human Sputum in Zambia: Public Health and Diagnostic Significance. *Infect. Dis. Ther.* **2013**, *1*, 1–4.

14.   Ayele, W.Y.; Neill, S.D.; Zinsstag, J.; Weiss, M.G.; Pavlik, I. Bovine tuberculosis: An old disease but a new threat to Africa. *Int. J. Tuberc. Lung Dis.* **2004**, *8*, 924–937. [PubMed]

15.   Etter, E.; Donado, P.; Jori, F.; Caron, A.; Goutard, F.L.; Roger, F. Risk Analysis and Bovine Tuberculosis, a Re-emerging Zoonosis. *Ann. N. Y. Acad. Sci.* **2006**, *1081*, 61–73. [CrossRef] [PubMed]

16.   Romha, G.; Gebre, G.; Ameni, G. Assessment of Bovine Tuberculosis and Its Risk Factors in Cattle and Humans, Assessment of bovine tuberculosis and its risk factors in cattle and humans, at and around Dilla town, southern Ethiopia. *Anim. Vet. Sci.* **2014**, *2*, 94–100. [CrossRef]

17.   Shitaye, J.E.; Tsegaye, W.; Pavlik, I. Bovine tuberculosis infection in animal and human populations in Ethiopia: A review. *Vet. Med.* **2007**, *52*, 317–332. [CrossRef]

18.   Sichewo, P.; Etter, E.; Michel, A. Prevalence of bovine tuberculosis in cattle at the wildlife/livestock/human interface in Northern Kwazulu-Natal province, South Africa. In Proceedings of the 9th Veterinary, Paraveterinary & Southern African Society for Veterinary Epidemiology and Preventive Medicine (SASVEPM) Congress, Boksburg, South Africa, 24–27 July 2017.

19.   Michel, A.; Coetzee, M.; Keet, D.; Mare, L.; Warren, R.; Cooper, D.; Bengis, R.; Kremer, K.; Van Helden, P. Molecular epidemiology of *Mycobacterium bovis* isolates from free-ranging wildlife in South African game reserves. *Veter. Microbiol.* **2009**, *133*, 335–343. [CrossRef] [PubMed]

20.   National Institute of Communicable Diseases SA. *Microbiologically Confirmed Tuberculosis 2004–15 South Africa*; National Institute of Communicable Diseases SA: Sandringham, Johannesburg, South Africa, 2016.

21.   Human Sciences Research Council (HSRC). *South African National HIV Prevalence, Incidence, Behaviour and Communication Survey, 2017*; Human Sciences Research Council: Pretoria, South Africa, 2018.

22.   TBFACTS.ORG. TB Statistics for South Africa—National & Provincial [Internet]. 2018. Available online: https://www.tbfacts.org/tb-statistics-south-africa/ (accessed on 12 March 2019).

23.   World Health Organisation. *Treatment of Tuberculosis Guidelines*; World Health Organisation: Geneva, Switzerland, 2010.

24. Palmer, M.V.; Waters, W.R. Advances in bovine tuberculosis diagnosis and pathogenesis: What policy makers need to know. *Vet. Microbiol.* **2006**, *112*, 181–190. [CrossRef] [PubMed]

25. Neeraja, D.; Veeregowda, B.M.; Rani, M.S.; Rathnamma, D.; Narayanaswamy, H.D.; Venkatesha, M.D.; Leena, G.; Apsana, R.; Somshekhar, S.H.; Saminathan, M.; et al. Identification of Mycobacterium tuberculosis Complex by Culture and Duplex Polymerase Chain Reaction in Bovines. *Asian J. Anim. Vet. Adv.* **2014**, *9*, 506–512.

26. Hardies, R.M.; Watson, J.M. *Mycobacterium bovis* in England and Wales: Past, present and future. *Epidemiol. Infect.* **1992**, *109*, 23–33.

27. De la Rua-Domenech, R. Human *Mycobacterium bovis* infection in the United Kingdom: Incidence, risks, control measures and review of the zoonotic aspects of bovine tuberculosis. *Tuberculosis* **2006**, *86*, 77–109. [CrossRef]

28. Tschopp, R.; Aseffa, A.; Schelling, E.; Berg, S.; Hailu, E.; Gadisa, E.; Zinsstag, J. Bovine tuberculosis at the wildlife-livestock-human interface in Hamer Woreda, South Omo, Southern Ethiopia. *PLoS ONE* **2010**, *5*, e12205. [CrossRef]

29. Roy, A.; Eisenhut, M.; Harris, R.J.; Rodrigues, L.C. Effect of BCG vaccination against Mycobacterium tuberculosis infection in children: Systematic review and meta-analysis. *Open Access* **2014**, *4643*, 1–11. [CrossRef] [PubMed]

30. Vayr, F.; Martin-Blondel, G.; Savall, F.; Soulat, J.-M.; Deffontaines, G.; Hérin, F. Occupational exposure to human *Mycobacterium bovis* infection: A systematic review. *PLoS Negl. Trop. Dis.* **2018**, *12*, e0006208. [CrossRef] [PubMed]

31. Debi, U.; Ravisankar, V.; Prasad, K.K.; Sinha, S.K.; Sharma, A.K. Abdominal tuberculosis of the gastrointestinal tract: Revisited. *World J. Gastroenterol.* **2014**, *20*, 14831–14840. [CrossRef] [PubMed]

32. Sharma, S.K.; Kohli, M.; Yadav, R.N.; Chaubey, J.; Bhasin, D.; Sreenivas, V.; Sharma, R.; Singh, B.K. Evaluating the Diagnostic Accuracy of Xpert MTB/RIF Assay in Pulmonary Tuberculosis. *PLoS ONE* **2015**, *10*, e0141011. [CrossRef] [PubMed]

33. Datta, S.; Shah, L.; Gilman, R.H.; Evans, A.C. Comparison of sputum collection methods for tuberculosis diagnosis: A systematic review and pairwise and network meta-analysis. *Lancet Glob. Heal.* **2017**, *5*, e760–e771. [CrossRef]

34. Katale, B.Z.; Mbugi, E.V.; Kendal, S.; Fyumagwa, R.D.; Kibiki, G.S.; Godfrey-Faussett, P.; Keyyu, J.D.; Van Helden, P.; Matee, M.I. Bovine tuberculosis at the human-livestock-wildlife interface: Is it a public health problem in Tanzania? A review. *Onderstepoort J. Vet. Res.* **2012**, *79*, 84–97. [CrossRef]

35. Rocha, A.; Elias, A.R.; Sobral, L.F.; Soares, D.F.; Santos, A.C.; Marsico, A.G.; Hacker, M.A.; Caldas, P.C.; Parente, L.C.; Silva, M.R.; et al. Genotyping did not evidence any contribution of *Mycobacterium bovis* to human tuberculosis in Brazil. *Tuberculosis* **2011**, *91*, 14–21. [CrossRef]

36. Ouassa, T.; Borroni, E.; Loukou, G.Y.; Faye-Kette, H.; Kouakou, J.; Menan, H.; Cirillo, D.M. High Prevalence of Shared International Type 53 among Mycobacterium Tuberculosis Complex Strains in Retreated Patients from Côte d'Ivoir. *PLoS ONE* **2012**, *7*, e45363. [CrossRef]

37. Adesokan, H.K.; Jenkins, A.O.; Soolingen, D.; Van Cadmus, S.I.B. *Mycobacterium bovis* infection in livestock workers in Ibadan, Nigeria: Evidence of occupational exposure. *Int. J. Tuberc. Lung Dis.* **2012**, *16*, 1388–1392. [CrossRef]

38. Fetene, T.; Kebede, N.; Alem, G. Tuberculosis Infection in Animal and Human Populations in Three Districts of Western Gojam, Ethiopia. *Zoonoses Public Heal.* **2011**, *58*, 47–53. [CrossRef]

39. Michel, A.L.; Geoghegan, C.; Hlokwe, T.; Raseleka, K.; Getz, W.M.; Marcotty, T. Longevity of *Mycobacterium bovis* in raw and traditional souring milk as a function of storage temperature and dose. *PLoS ONE* **2015**, *10*, e0129926. [CrossRef] [PubMed]

40. Mazwi, K.D. Longitudinal Monitoring of Mycobacterium bovis and Non-Tuberculous Mycobacteria in Cow's Milk from a Rural Community. Dissertation's Thesis, Department of Veterinary Tropical Disease, University of Pretoria, Pretoria, South Africa, 2018.

41. Rodwell, T.C.; Moore, M.; Moser, K.S.; Brodine, S.K.; Strathdee, S.A. Tuberculosis from *Mycobacterium bovis* in Binational Communities, United States. *Emerg. Infect. Dis.* **2008**, *14*, 909–916. [CrossRef] [PubMed]

42. Mfinanga, S.G.; Mørkve, O.; Kazwala, R.R.; Cleaveland, S.; Sharp, J.M.; Shirima, G.; Nilsen, R. The role of livestock keeping in tuberculosis trends in Arusha, Tanzania. *Int. J. Tuberc. Lung Dis.* **2003**, *7*, 695–704.

43. Ibrahim, S.; Cadmus, S.I.B.; Umoh, J.U.; Ajogi, I.; Farouk, U.M.; Abubakar, U.B.; Kudi, A.C. Tuberculosis in humans and cattle in Jigawa State, Nigeria: Risk factors analysis. *Vet. Med. Int.* **2012**, *2012*, 865924. [CrossRef] [PubMed]

44. Nafarnda, W.D.; Obudu, C.E.; Omeiza, G.K.; Enem, S.I.; Adeiza, M.A. Prevalence of Zoonotic bovine tuberculosis and associated risk factors among cattle herds in North Central Nigeria. *Int. J. Curr. Res. Acad. Rev.* **2015**, *3*, 113–119.

45. Mbugi, E.V.; Katale, B.Z.; Siame, K.K.; Keyyu, J.D.; Kendall, S.L.; Dockrell, H.M.; Streicher, E.M.; Michel, A.L.; Rweyemamu, M.M.; Warren, R.M.; et al. Genetic diversity of Mycobacterium tuberculosis isolated from tuberculosis patients in the Serengeti ecosystem in Tanzania. *Tuberculosis* **2015**, *95*, 170–178. [CrossRef] [PubMed]

46. Adesokan, H.K.; Akinseye, V.O.; Sulaimon, M.A. Knowledge and practices about zoonotic tuberculosis prevention and associated determinants amongst livestock workers in Nigeria; 2015. *PLoS ONE* **2018**, *13*, e0198810. [CrossRef]

47. Warren, R.M.; Van Pittius, N.C.G.; Barnard, M.; Hesseling, A.; Engelke, E.; De Kock, M.; Gutierrez, M.C.; Chege, G.K.; Victor, T.C.; Hoal, E.G.; et al. Differentiation of Mycobacterium tuberculosis complex by PCR amplification of genomic regions of difference. *Int. J. Tuberc. Lung Dis.* **2006**, *10*, 818–822.

48. Gcebe, N.; Rutten, V.; Gey van Pittius, N.C.; Michel, A. Prevalence and distribution of non-tuberculous mycobacteria (NTM) in cattle, African buffaloes (*Syncerus caffer*) and their environments in South Africa. *Transbound. Emerg. Dis.* **2013**, *60* (Suppl. 1), 74–84. [CrossRef]

49. Hlokwe, T.M.; van Helden, P.; Michel, A. Evaluation of the Discriminatory Power of Variable Number of Tandem Repeat Typing of *Mycobacterium bovis* Isolates from Southern Africa. *Transbound. Emerg. Dis.* **2013**, *60*, 111–120. [CrossRef]

50. Venables, W.N.; Ripley, B.D. *Modern Applied Statistics with S*; Statistics and Computing; Springer: New York, NY, USA, 2002.

 *pathogens* 

Communication

Serological and Molecular Investigation on *Toxoplasma gondii* Infection in Wild Birds

Simona Nardoni [1],* , Guido Rocchigiani [1], Ilaria Varvaro [1], Iolanda Altomonte [1], Renato Ceccherelli [2] and Francesca Mancianti [1]

[1] Dipartimento di Scienze Veterinarie, Università degli Studi di Pisa, 56124 Pisa, Italy; guido.rocchigiani.g@gmail.com (G.R.); ilaria5787@virgilio.it (I.V.); altomonte@vet.unipi.it (I.A.); francesca.mancianti@unipi.it (F.M.)

[2] Centro Recupero Uccelli Marini e Acquatici-CRUMA, 57121 Livorno, Italy; apusvet.cruma@libero.it

* Correspondence: simona.nardoni@unipi.it; Tel.: +39-050-2216950

Received: 12 April 2019; Accepted: 24 April 2019; Published: 29 April 2019

Abstract: *Toxoplasma gondii* is an obligate apicomplexan zoonotic parasite that infects humans and other animals and is responsible for toxoplasmosis. This parasite causes one of the most common parasitic infections in humans worldwide. Toxoplasmosis meets the requirements for a One Health Disease due to its ability to affect the health of human beings as well as domestic and free ranging animals. Integrating human, domestic animal, and wildlife data could better assess the risk and devise methods of control. A first step of such an approach would be the knowledge of the prevalence of parasitosis in humans and animals in selected areas. Therefore, the aim of the present study was to evaluate the occurrence of *Toxoplasma* infection in 216 free ranging birds belonging to different genera/species by serology and molecular techniques. Twenty-five out of 216 animals (11.6%) were positive to the immunofluorescence antibody test (IFAT) with antibody titers ranging from 1/20 to 1/320, and 19 of them (8.8%) also showed a positive PCR for *Toxoplasma* DNA. The results confirmed the widespread occurrence of *Toxoplasma* infection in wild birds and serological data were corroborated by molecular results in birds that also had low antibody titers. The knowledge of the wide occurrence of the parasite in game and wild birds should enhance the accurate estimation of the risks in handling, managing, and eating these species with regard to domestic carnivores as well as the impact of viscera and offal in the environment.

Keywords: *Toxoplasma gondii*; birds; IFAT; serology; PCR; zoonosis; One Health

---

1. Introduction

Toxoplasma gondii is an obligate apicomplexan zoonotic parasite that infects humans and other animals and is responsible for toxoplasmosis. The life cycle of this intracellular parasite is very complex and encounters two stages. The sexual stage develops in the intestine of wild and domestic felid hosts, while the asexual phase takes place in all warm-blooded animals including birds and mammals.

Animal infection can take place by the ingestion of small mammals and birds bearing tissue cysts and/or of oocysts shed by the feline final host and sporulated in the environment. Horizontal transmission in humans follows the oral intake of raw or undercooked meat as well as food and water contaminated by sporulated oocysts.

The parasite is widespread throughout the world, and toxoplasmosis represents one of the most common parasitic infections in humans. Although the infection is most typically asymptomatic in immunocompetent subjects, a number of reports about the occurrence of ocular symptoms such as retinochoroiditis and retinitis consequent to acquired toxoplasmosis in humans are present in the literature. Immunocompromised patients are at risk from severe disease following both primary infection and reactivated toxoplasmosis [1]. If primary infection is contracted during pregnancy, intrauterine infection may occur in immunocompetent women, with transmission to the fetus [2,3].

Toxoplasmosis meets the requirements for a One Health Disease due to its ability to affect the health of human beings as well as domestic and free ranging animals. It also impacts on ecosystems and is a threat to those who rely on animal resources [4–6]. Recently, the One Health approach to toxoplasmosis has been excellently revised by Aguirre et al. [6], hoping for transdisciplinary collaborations, by monitoring toxoplasmosis and *T. gondii* prevalence. Integrating human, domestic animal, and wildlife data could better assess the risk and devise methods of control. A first step of such an approach would be acquiring the knowledge of the prevalence of parasitosis in human and animals in selected areas.

Toxoplasmosis in wildlife such as many other parasite zoonoses has little clinical impact, however, the spillover from wildlife to human and/or domestic animals should be considered [7].

Avian species are usually resistant to *Toxoplasma* infection, even if pigeons and canaries can be severely affected [8]. Wild birds are important in the *T. gondii* epidemiology, which considered their role as a reservoir for carnivores. Some of them also have migrating behavior and could spread the parasite worldwide.

Data about the occurrence of *Toxoplasma* infection in wild birds in Italy are scant and to the best of our knowledge, there have only been two reports referring to waterfowl and birds of prey, respectively [9,10]. For these reasons, the aim of the present study was to evaluate the occurrence of *Toxoplasma* infection in various free ranging bird species by serology and molecular techniques.

2. Results

Twenty five out of 216 animals (11.6%) were positive to the immunofluorescence antibody test (IFAT) with antibody titers ranging from 1/20 to 1/320, and 19 of them (8.8%) also showed a positive PCR for *Toxoplasma* DNA. In detail, DNA was found in the hearts of 11 birds, in the brains of four animals, while the other four subjects had parasite DNA both in the heart and brain.

Detailed results are reported in Table 1.

Table 1. Species, gender, and the results of the serology and PCR on positive birds.

Bird Species	Gender	IFAT	PCR Heart	PCR Brain
Anas crecca	male	1/40	negative	positive
Anas crecca	female	1/20	positive	negative
Anas crecca	female	1/20	negative	negative
Anas crecca	female	1/80	positive	negative
Anas crecca	female	1/40	positive	negative
Anas crecca	female	1/20	negative	negative
Anas crecca	female	1/20	negative	negative
Anas crecca	female	1/80	negative	positive
Anas penelope	female	1/40	positive	negative
Anas penelope	female	1/20	positive	positive
Anas plathyrinchos	male	1/20	positive	negative
Anas plathyrinchos	female	1/40	negative	negative
Buteo buteo	male	1/160	negative	negative
Columba palumbus	female	1/80	positive	positive
Falcus tinnunculus	male	1/160	negative	negative
Falcus tinnunculus	male	1/320	positive	positive
Falcus tinnunculus	female	1/320	positive	positive

Table 1. *Cont.*

Bird Species	Gender	IFAT	PCR Heart	PCR Brain
Gallinago gallinago	female	1/40	negative	negative
Larus ridibundus	male	1/160	positive	negative
Larus ridibundus	female	1/160	negative	positive
Larus ridibundus	female	1/320	positive	negative
Larus ridibundus	male	1/160	positive	negative
Phasianus colchicus	male	1/80	positive	negative
Sturnus vulgaris	male	1/80	positive	negative
Vanellus vanellus	male	1/80	positive	negative

Twelve hunted waterfowl out of 148 (8.1%) were seropositive and eight of them (5.4%) harbored parasite DNA, while the birds from a rescue center yielded a seroprevalence of 19.1% and 13.2% gave a positive PCR result.

3. Discussion

In the present report, an overall seroprevalence of 11.6% and positive PCR of 8.8% were found. Although the selected bird population was very heterogeneous and the number of subjects was low, these preliminary data are of interest. There have only been a few studies reporting on the occurrence of antibodies or DNA in avian species in Europe, except for Spain. To the best of our knowledge, information about the simultaneous presence of the two data is lacking, except for a previous study carried out by us on waterfowl from Tuscany [9]. In this paper, an 8.7% seroprevalence along with a PCR positivity of 2.9% was reported. The data referring to waterfowl in the present study gave a similar seroprevalence, but the occurrence of parasite DNA was more than twice. Moreover, surprisingly, in the present study, the heart showed a greater sensitivity, yielding positive results for parasite DNA in 15 animals, while the brains were positive in only eight subjects.

Serological studies from Europe regarding raptors showed a prevalence of 50% from Portugal [9], from 0% to 79% in France [11,12] and 10.7% from Italy [10]. Other studies have reported seroprevalence values of 80.5% in common ravens [13], and 26.1% in various avian species from Spain [14]. Molecular studies have been carried out on the brains of different avian species in Spain, with an overall PCR positivity of 6% [15].

The results from this preliminary report would fit with data referred by other countries, and four raptors out of 15 (27%) were seropositive. Four out of 18 seagulls (22.3%) scored seropositive, in full agreement with Cabezon et al. [16] who first reported the occurrence of antibodies (21%) versus *T. gondii* in 525 seagull chicks, and with Gamble et al. [17], who reported a large exposure of yellow-legged gull to the parasite by checking 1122 eggs and hypothesized of their involvement in the maintenance and circulation of toxoplasma, suggesting that large gulls could be used as epidemiological sentinels at the human–wildlife interface.

These results confirm the widespread occurrence of toxoplasma infection in wild birds and the serological data were also corroborated by molecular results in waterfowl with low antibody titers (1/20).

One Health has emphasized the need to bridge disciplines linking human health, animal health, and ecosystem health [4]. Human activities can influence the zoonotic transmission of *Toxoplasma* involving wildlife. Hunting, the lack of control of domestic hosts (diet and roaming), dietary human factors, and environmental contamination are the most important risk factors [7]. The One Health triad encompasses the collaborative goals of providing optimal health for people, animals, and the environment by considering interactions between all three systems [7]. Therefore, knowledge of the wide occurrence of the parasite in game and wild birds should enhance an accurate estimation of the risks in handling, managing, and eating these species also with regard to domestic carnivores as well as to the impact of viscera and offal into the environment.

4. Material and Methods

The study was carried out on n = 216 birds (Table 2) belonging to different avian species, referred to the Veterinary Parasitology Lab for epidemiological purposes. Part of them (n = 148) were waterfowl from regular hunting activity. The other birds (n = 68) were referred by a rescue center for avian species (CRUMA) and were injured animals, or who had died during hospitalization.

Table 2. Number, species and gender of birds studied.

Species	Total Number	Males	Females
Anas crecca	73	23	50
Anas platyrhyncos	28	15	13
Anas penelope	23	10	13
Anas acuta	5	0	5
Anas strepera	3	0	2
Aythya ferina	3	2	1
Tadorna tadorna	3	1	2
Anas querquendula	1	0	1
Aythya fuligula	1	0	1
Anas clypeata	8	2	6
Anser anser	1	1	1
Sturnus vulgaris	2	2	0
Corvus frugilegus	2	1	1
Sylvia melanocephala	1	1	0
Garrus glandarius	1	0	1
Spinus spinus	1	1	0
Parus major	1	1	0
Phylloscopus collybita	1	0	1
Columba livia	3	0	3
Columba palumbus	8	6	2
Falco tinnunculus	6	4	2
Falcus peregrinus	1	1	0
Phasianus colchicus	3	2	1
Gallinago gallinago	6	3	3
Vanellus vanellus	2	2	0
Larus ridibundus	18	4	14
Buteo buteo	4	2	2
Asio otus	1	1	0
Fulica atra	2	2	0
Athene noctua	3	1	2
Ardea cinerea	1	1	0

The whole brain and heart [18] as well as intracardiac coagules were collected from all birds, shortly after their delivery at the Lab.

Tissues were kept at −20 °C until processing for the molecular detection of *Toxoplasma* DNA, and serum was employed to evaluate the presence of anti-*Toxoplasma* antibodies by IFAT, using 12-well slides (Toxospot ®, BioMérieux, Marcy l'Etoile, France) and an anti-chicken IgG FITC (Sigma-Aldrich s.r.l., Milan, Italy) diluted 1/32. This test was preferred to MAT, considering that most sera were hemolyzed, making it hard to read the results. Therefore, IFAT was employed, as reported by Maksimov et al. [19], with slight modification, starting from a threshold dilution of 1/20.

DNA was extracted from about 200 mg of homogenized tissues to perform PCR analysis for *Toxoplasma* DNA using the DNeasy® Blood & Tissue (Qiagen, Milan, Italy) in accordance with the manufacturer's instructions. After extraction, DNA was stored at −20 °C until use.

Two pairs of oligonucleotide primers were used to amplify regions of the B1 gene of *T. gondii*: the outer primers 5′-GGAACTGCATCCGTTCATGAG-3′ (sense strand) and 5′-TCTTTAAAGCGTTCGTG GTC-3′ (nonsense strand) and inner primers 5′-TGCATAGGTTGCAGTCACTG-3′ (sense strand) and 5′-GGCGACCAATCTGCGAATACACC-3′ (nonsense strand), provided by Eurofins MGW (M-Medical, Milano, Italy). Nested PCR was performed as described by Jones et al. [20].

Author Contributions: Conceptualization, F.M. and S.N.; Methodology, F.M. and S.N.; Formal analysis, R.C., G.R., I.V., and I.A.; Data curation, F.M. and S.N.; Writing—original draft preparation, F.M. and S.N.; Writing—review and editing, F.M. and S.N.

Funding: This research received no external funding.

Conflicts of Interest: The authors declare no conflict of interest.

References

1. Khan, K.; Khan, W. Toxoplasmosis: An overview of the neurological and ocular manifestations. *Parasitol. Int.* **2018**, *67*, 715–721. [CrossRef] [PubMed]

2. Tenter, A.M.; Heckeroth, A.R.; Weiss, L.M. *Toxoplasma gondii*: From animals to humans. *Int. J. Parasitol.* **2001**, *30*, 1217–1258. [CrossRef]

3. Fallahi, S.; Rostami, A.; Nourollahpour Shiadeh, M.; Behniafar, H.; Paktinat, S. An updated literature review on maternal-fetal and reproductive disorders of *Toxoplasma gondii* infection. *J. Gynecol. Obstet. Hum. Reprod.* **2018**, *47*, 133–140. [CrossRef] [PubMed]

4. Crozier, G.; Schulte-Hostedde, A.I. The ethical dimensions of wildlife disease management in an evolutionary context. *Evol. Appl.* **2014**, *7*, 788–798. [CrossRef] [PubMed]

5. Jenkins, E.J.; Simon, A.; Bachand, N.; Stephen, C. Wildlife parasites in a One Health world. *Trends Parasitol.* **2015**, *31*, 174–180. [CrossRef] [PubMed]

6. Aguirre, A.A.; Longcore, T.; Barbieri, M.; Dabritz, H.; Hill, D.; Klein, P.N.; Lepczyk, C.; Lilly, E.L.; McLeod, R.; Milcarsky, J.; et al. The One Health Approach to Toxoplasmosis: Epidemiology, Control, and Prevention Strategies. *Ecol. Health* **2019**, *3*, 1–3. [CrossRef] [PubMed]

7. Thompson, R.C. Parasite zoonoses and wildlife: One Health, spillover and human activity. *Int. J. Parasitol.* **2013**, *43*, 1079–1088. [CrossRef] [PubMed]

8. Dubey, J.P. A review of toxoplasmosis in wild birds. *Vet. Parasitol.* **2002**, *106*, 121–153. [CrossRef]

9. Mancianti, F.; Nardoni, S.; Mugnaini, L.; Poli, A. *Toxoplasma gondii* in waterfowl: The first detection of this parasite in *Anas crecca* and *Anas clypeata* from Italy. *J. Parasitol.* **2013**, *99*, 561–563. [CrossRef] [PubMed]

10. Gazzonis, A.L.; Zanzani, S.A.; Santoro, A.; Veronesi, F.; Olivieri, E.; Villa, L.; Lubian, E.; Lovati, S.; Bottura, F.; Epis, S.; et al. *Toxoplasma gondii* infection in raptors from Italy: Seroepidemiology and risk factors analysis. *Comp. Immunol. Microbiol. Infect. Dis.* **2018**, *60*, 42–45. [CrossRef] [PubMed]

11. Lopes, A.P.; Sargo, R.; Rodrigues, M.; Cardoso, L. High seroprevalence of antibodies to *Toxoplasma gondii* in wild animals from Portugal. *Parasitol. Res.* **2011**, *108*, 1163–1169. [CrossRef] [PubMed]

12. Aubert, D.; Terrier, M.E.; Dumètre, A.; Barrat, J.; Villena, I. Prevalence of *Toxoplasma gondii* in raptors from France. *J. Wildl. Dis.* **2008**, *44*, 172–173. [CrossRef] [PubMed]

13. Molina-López, R.; Cabezón, O.; Pabón, M.; Darwich, L.; Obón, E.; Lopez-Gatius, F.; Dubey, J.P.; Almería, S. High seroprevalence of *Toxoplasma gondii* and *Neospora caninum* in the Common raven (*Corvus corax*) in the Northeast of Spain. *Res. Vet. Sci.* **2012**, *93*, 300–302. [CrossRef] [PubMed]

14. Cabezón, O.; García-Bocanegra, I.; Molina-López, R.; Marco, I.; Blanco, J.M.; Höfle, U.; Margalida, A.; Bach-Raich, E.; Darwich, L.; Echeverría, I.; et al. Seropositivity and risk factors associated with *Toxoplasma gondii* infection in wild birds from Spain. *PLoS ONE* **2011**, *6*, e29549. [CrossRef]

15. Darwich, L.; Cabezón, O.; Echeverria, I.; Pabón, M.; Marco, I.; Molina-López, R.; Alarcia-Alejos, O.; López-Gatius, F.; Lavín, S.; Almería, S. Presence of *Toxoplasma gondii* and *Neospora caninum* DNA in the brain of wild birds. *Vet. Parasitol.* **2012**, *183*, 377–381. [CrossRef] [PubMed]

16. Cabezón, O.; Cerdà-Cuéllar, M.; Morera, V.; García-Bocanegra, I.; González-Solís, J.; Napp, S.; Ribas, M.P.; Blanch-Lázaro, B.; Fernández-Aguilar, X.; Antilles, N.; et al. *Toxoplasma gondii* infection in seagull chicks is related to the consumption of freshwater food resources. *PLoS ONE* **2016**, *11*, e0150249. [CrossRef]

17. Gamble, A.; Ramos, R.; Parra-Torres, Y.; Mercier, A.; Galal, L.; Pearce-Duvet, J.; Villena, I.; Montalvo, T.; González-Solís, J.; Hammouda, A.; et al. Exposure of yellow-legged gulls to *Toxoplasma gondii* along the Western Mediterranean coasts: Tales from a sentinel. *Int. J. Parasitol. Parasites Wildl.* **2019**, *8*, 221–228. [CrossRef] [PubMed]

18. Alvarado-Esquivel, C.; Rajendran, C.; Ferreira, L.R.; Kwok, O.C.H.; Choudhary, S.; Alvarado-Esquivel, S.; Rodríguez-Peña, S.; Villena, I.; Dubey, J.P. Prevalence of *Toxoplasma gondii* infection in wild birds in Durango, Mexico. *J. Parasitol.* **2011**, *97*, 809–812. [CrossRef] [PubMed]

19. Maksimov, P.; Buschtöns, S.; Herrmann, D.C.; Conraths, F.J.; Görlich, K.; Tenter, A.M.; Dubey, J.P.; Nagel-Kohl, U.; Thoms, B.; Bötcher, L.; et al. Serological survey and risk factors for *Toxoplasma gondii* in domestic ducks and geese in Lower Saxony, Germany. *Vet. Parasitol.* **2011**, *182*, 140–149. [CrossRef] [PubMed]

20. Jones, C.D.; Okhravi, N.; Adamson, P.; Tasker, S.; Lightman, S. Comparison of PCR Detection Methods for B1, P30, and 18s rDNA Genes of *T.gondii* in Aqueous Humor. *Investig. Ophthalmol. Vis. Sci.* **2000**, *41*, 634–644.

 *pathogens*

Article

Serological Evidence of *Anaplasma phagocytophilum* and Spotted Fever Group *Rickettsia* spp. Exposure in Horses from Central Italy

Valentina Virginia Ebani [1,2]

1   Department of Veterinary Science, University of Pisa, viale delle Piagge 2, 56124 Pisa, Italy;
    valentina.virginia.ebani@unipi.it
2   CIRSEC, Center for Climatic Change Impact, University of Pisa, via del Borghetto 80, 56124 Pisa, Italy

Received: 22 May 2019; Accepted: 24 June 2019; Published: 26 June 2019

Abstract: *Anaplasma phagocytophilum* and *Rickettsia* spp. are tick-borne bacteria of veterinary and human concern. In view of the One-Health concept, the present study wanted to evaluate the spreading of these pathogens in horses living in central Italy. In particular, the aim of the investigation was to verify the exposure to *A. phagocytophilum* in order to update the prevalence of this pathogen in the equine population from this area, and to spotted fever group (SFG) *Rickettsia* spp. to evaluate a possible role of horses in the epidemiology of rickettsiosis. Indirect immunofluorescent assay was carried out to detect antibodies against *A. phagocytophilum* and SFG (spotted fever group) *Rickettsia* spp. in blood serum samples collected from 479 grazing horses living in central Italy during the period from 2013 to 2018. One hundred and nine (22.75%) horses were positive for *A. phagocytophilum*, 72 (15.03%) for SFG *Rickettsia* spp., and 19 (3.96%) for both antigens. The obtained results confirm the occurrence of *A. phagocytophilum* in equine populations, and also suggest the involvement of horses in the epidemiology of SFG rickettsiosis. In both cases, in view of the zoonotic aspect of these pathogens and the frequent contact between horses and humans, the monitoring of equine populations could be useful for indication about the spreading of the tick-borne pathogens in a certain geographic area.

Keywords: horses; *Anaplasma phagocytophilum*; spotted fever group *Rickettsia* spp.; zoonosis; tick-borne infections

---

1. Introduction

Hematophagous arthropods, especially ticks, are well-known as vectors of several bacterial, viral, and protozoan pathogens. Among them, *Anaplasma phagocytophilum* and *Rickettsia* spp. may induce clinical manifestations in humans and different animal species.

Anaplasma phagocytophilum is an obligate intracellular, Gram negative bacterium belonging to the order *Rickettsiales*, family *Anaplasmataceae*. It is able to infect granulocytes, mainly neutrophils, of several domestic and wild animal species [1].

Wild mammals usually serve as asymptomatic reservoirs of *A. phagocytophilum*, whereas domestic animals may develop clinically defined diseases, such as tick-borne fever in cattle and sheep and granulocytic anaplasmosis in dogs [2]. Furthermore, *A. phagocytophilum* is cause of infection in human beings, who develop a disease called human granulocytic anaplasmosis (HGA), varying from mild to severe forms with fever, headache, myalgia, arthralgia, leukopenia, and thrombocytopenia; moreover, serious opportunistic infections can occur in immunocompromised patients during the course of HGA [3].

Horses infected by this pathogen develop a disease known as equine granulocytic anaplasmosis (EGA) (formerly equine granulocytic ehrlichiosis), characterized by a wide range of clinical signs. Usually, infected horses show fever, lethargy, ataxia, reluctance to move, icterus, and petechiation; laboratory blood abnormalities may include leukopenia, thrombocytopenia, and anemia [4].

In all animal species, *A. phagocytophilum* is transmitted by a tick bite. *Ixodes* species are involved in the epidemiology of this pathogen worldwide, in particular *I. ricinus* in Europe, including Italy [2].

The genus *Rickettsia* comprises obligate intracellular, Gram-negative bacteria transmitted by hematophagous arthropods. Spotted fever group (SFG) includes several *Rickettsia* species responsible for disease, often serious, in animals and humans. *Rickettsia conorii* is the etiologic agent of the Mediterranean spotted fever (MSF) that represents the most widespread SFG rickettiosis in the Mediterranean countries, including Italy, especially in the southern (Sardinia, Sicily, Calabria) and central regions [5], where most cases have been reported mainly between the months of June and September [6].

In Italy, several SFG rickettsiae are circulating, as mainly demonstrated by molecular investigations on tick populations. In particular, DNA of *R. conorii*, *R. helvetica*, *R. massiliae*, *R. slovaca*, *R. monacensis*, *R. aeschlimannii*, *R. raoultii*, and *R. africae* have been detected [7–12].

No data about the spreading of rickettsiae among horses living in Italy is available, and very scant information about equine rickettsiosis in other countries is present in the scientific literature [13,14].

Considering that anaplasmosis and rickettsiosis are zoonotic infections and the high occasion of contact between humans and horses, in view of the One-Health concept, the present study wanted to evaluate the spreading of these pathogens in horses living in central Italy. In particular, the aim of the investigation was to verify the exposure to *A. phagocytophilum* in order to update the prevalence of this pathogen in equine population from this area, and to SFG *Rickettsia* spp., to evaluate a possible role of horses in the epidemiology of rickettsiosis.

2. Results

Among the 479 tested horses, 109 resulted positive for *A. phagocytophilum*, with 22.75% total mean seroprevalence; prevalence values observed in the different years varied from 17.46% (2013) to 27.08% (2018). Antibody titers ranged from 1:40 to 1:1280.

Seventy two (15.03%) horses had antibodies to SFG *Rickettsia* spp. Prevalence values ranged from 11.26% (2016) to 17.71% (2018) in relation with the years of sampling. Furthermore, antibody titers from 1:64 to 1:1024 were observed. Nineteen (3.96%) horses had antibodies to both *A. phagocytophilum* and SFG *Rickettsia* spp. Results are summarized in Tables 1–3.

Table 1. Results obtained by indirect immunofluorescence test for *Anaplasma phagocytophilum* in relation to years of sampling and antibody titers.

Years	N. Tested Horses	N. Positive (%) Horses	Antibody Titers (%)					
			1:40	1:80	1:160	1:320	1:640	1:1280
2013	63	11 (17.46)	2	7	1	1	-	-
2014	76	15 (19.73)	3	8	3	-	1	-
2015	92	22 (23.91)	5	13	2	1	-	1
2016	71	14 (19.71)	3	8	1	2	-	-
2017	81	21 (25.92)	4	15	1	-	-	1
2018	96	26 (27.08)	7	11	5	1	2	-
Total	479	109 (22.75)	24 (5.01)	62 (12.94)	13 (2.71)	5 (1.04)	3 (0.63)	2 (0.42)

Table 2. Results obtained by indirect immunofluorescence test for spotted fever group (SFG) *Rickettsia* spp. in relation to years of sampling and antibody titers.

Years	N. Tested Horses	N. Positive (%) Horses	Antibody Titers (%)				
			1:64	1:128	1:256	1:512	1:1024
2013	63	9 (14.28)	5	3	-	1	-
2014	76	13 (17.10)	4	5	3	-	1
2015	92	11 (11.95)	6	5	-	-	-
2016	71	8 (11.26)	3	3	2	-	-
2017	81	14 (17.28)	4	2	5	3	
2018	96	17 (17.71)	8	6	2	-	1
Total	479	72 (15.03%)	30 (6.26)	24 (5.01)	12 (2.50)	4 (0.84)	2 (0.42)

Table 3. Horses resulted positive to both *Anaplasma phagocytophilum* and SFG *Rickettsia* spp. with indirect immunofluorescence test.

Horses	*Anaplasma phagocytophilum*	SFG *Rickettsia* spp.
1	1:80	1:64
2	1:80	1:64
3	1:80	1:128
4	1:40	1:64
5	1:160	1:64
6	1:40	1:256
7	1:80	1:128
8	1:160	1:128
9	1:80	1:128
10	1:80	1:64
11	1:80	1:64
12	1:80	1:128
13	1:40	1:128
14	1:40	1:64
15	1:320	1:64
16	1:40	1:64
17	1:40	1:256
18	1:80	1:128
19	1:80	1:64

3. Discussion

Tick-borne diseases are major animal and human health issues in several geographic areas. Global warming has deeply influenced the spread of hematophagous arthropods, including ticks, but other factors are involved in their distribution. Animals' movements, agricultural and wildlife management, and urbanization with consequent reduction of natural areas have determined changes in tick distribution, with increasing presence in urban and peri-urban environment.

Horses examined in this study lived in areas with environmental conditions which favor tick diffusion, especially abundant vegetation and presence of other animal species, mainly wildlife. Several tick species are present in this area, including *Ixodes ricinus*, which is the main vector of *A. phaocytophilum* [2] and is also involved in the transmission of rickettiae. Furthermore, the brown dog tick *Rhipicephalus sanguineus*, the main vector of *R. conori*, may be found not only among dogs, but also in wildlife [15].

Results obtained during this survey showed that the examined horses had been exposed to the investigated tick-borne pathogens, with a higher seroprevalence detected for *A. phagocytophilum* than for SFG *Rickettsia* spp. Higher values of seroprevalence for both pathogens were detected among samples collected in 2017 and 2018. These results could be related to climatic conditions that allowed a higher presence of ticks in the areas where the tested horses lived. However, the spread of arthropods is related to further factors (presence of other animals, environmental management, acaricide treatments) that were not fully known in this study.

Previous investigations found *A. phagocytophilum* infection in equine populations in Italy: in detail, serological surveys detected prevalence values ranging between 9% and 17%, whereas molecular studies found prevalences from 4.7% to 25.62% [16–20]. Seroprevalences found in Europe, which ranged from 11.3% to 20%, were quite similar [21–23]; conversely, molecular surveys carried out in some European countries detected lower prevalences that varied from 1.4% to 9.8% [24–26].

Furthermore, *A. phagocytophilum* infection was detected in other animal species living in Italy [27–30], as well as *A. phagocytophilum* DNA being found in ticks collected from animals or environment [31]. The present results confirm the spreading of this tick-borne pathogen among equine populations in the investigated areas. Moreover, considering that the tested animals did not show clinical signs, the results corroborated that asymptomatic forms may be developed by infected horses. In fact, some authors affirm that horses from endemic areas have a higher seroprevalence to *A. phagocytophilum* than those from non-endemic areas, and horses introduced into an endemic area are more likely to develop illness than native horses [32]. However, clinical findings, when present, are not specific, and it can be difficult to differentiate EGA from other diseases, mainly piroplasmosis. Moreover, horses infected by *A. phagocytophilum* are predisposed to developing secondary infections, which may complicate the clinical diagnosis [32].

Data about the presence of SFG *Rickettsia* spp. in Italy mainly concern humans [33]. Some studies have been carried out in ticks and wild animals, and case reports of canine rickettsiosis have also been documented [34,35]. Data about rickettsiosis in Italian equine populations are not available in the scientific literature, and very scant studies have been reported from other European countries. Elfving et al. [13] tested sera from 63 horses in Sweden with indirect immunofluorescence assay (IFA) employing *R. helvetica* as antigen, and found a 36.5% prevalence; Skotarczack et al. [14] found *R. helvetica* and *R. monacensis* DNA in two ticks collected from ponies in Poland.

Furthermore, serological surveys have been carried out in horses living in South America. A 2.85% seroprevalence for SFG *Rickettsia* spp. was found in horses from Colombian Orinoquia [36]; in Brazil, 183/504 (36.3%) horses were seropositive for *Rickettsia rickettsii* [37], whereas among 258 horses tested for *R. rickettsii*, *R. amblyommatis*, and *R. bellii*, 152 (58.91%) were seroreactive for at least one *Rickettsia* species [38].

To the best of our knowledge, this is the first serological survey on the occurrence of SFG *Rickettsia* spp. in horses from Italy. Finding *Rickettsia*-positive horses suggests that they can contribute to the natural cycle of these bacteria as hosts for infected ticks. Equine illness related to rickettsia has never been described, therefore horses could respond immunologically to exposure to rickettsiae without developing clinical signs.

An experimental infection of horses with *R. rickettsii* demonstrated that the animals had no bacteremia, clinical, hematological, or blood biochemical alterations, but they had specific antibodies from 10 days to 2 years after infection [39].

During the present investigation, *R. conorii*, which belongs to the SFG, was used as an antigen for IFA. This pathogen, an agent of MSF disease, has been frequently found in Italy, as demonstrated by serological and molecular studies [33,35,40,41].

Horses resulting positive for rickettsiosis could have antibodies to *R. conorii*, as well as to other SFG *Rickettsia* spp. present in Italy, because IFA, even though considered the gold standard method for the serological diagnosis of rickettsiosis [42], is not able to differentiate between antibodies against the different SFG species [5].

Horses are largely employed in agonistic activity and are frequently maintained for recreational purposes, therefore, humans are highly exposed to the risk of being bitten by ticks previously fed on infected horses, as suggested by some authors who have considered human contact with horses as a risk factor for acquiring tick-borne lymphadenopathy (TIBOLA) caused by *R. slovaca* [43].

Detection of horses with antibodies to both *A. phagocytophilum* and SFG *Rickettsia* spp. confirms that the equine population, as well as human beings and other animals, may be affected by more arthropod-borne pathogens as consequence of the transmission by one tick harboring different microorganisms and/or more infected ticks. *A. phagocytophilum*, *Borrelia burgdorferi* sensu lato, *Coxiella burnetii*, and piroplasms have been demonstrated to be responsible for co-infections in clinically healthy and symptomatic horses [19,20,22]. Further studies could be useful to investigate if rickettsiae may complicate equine clinical forms when involved in co-infections.

4. Material and Methods

4.1. Animals

From January 2013 to December 2018, peripheral whole blood samples were collected from 479 grazing horses. Animals were actively racing and lived in various farms and horse centres located in lowland and hilly areas of Central Italy; they did not show clinical signs and were not under antibiotic treatment. Breeders and owners reported previous tick exposure.

Whole blood samples (about 10 mL), drawn from the right or left jugular vein, were centrifuged at 1500× *g* for 15 min, and the sera were collected and immediately tested or stored at −20 °C.

Ethical Statement

The collection of blood samples was executed for other clinical exams as part of routine health care by collaborating veterinarians during clinical visits. All animals were treated with standard practices of animal care and no horses were submitted to blood collection for this study only. However, in all cases, informed consent was obtained from the owners.

4.2. Serological Analyses

The indirect immunofluorescence antibody test (IFA) was executed on IFA slides prepared with *Anaplasma phagocytophilum* and *Rickettsia conorii* (Fuller Laboratories Fullerton, California, USA) antigens, respectively.

Blood sera were diluted 1:40 and 1:64 in phosphate-buffered saline (PBS, pH 7.2), considered cut-off values for *A. phagocytophilum* and *Rickettsia* spp., respectively, as reported in previous studies [13,44]. The test was executed employing a rabbit fluorescein isothiocyanate-conjugated anti-horse IgG (Sigma-Aldrich, Milano, Italy) diluted 1:30 in Evans Blue (Sigma-Aldrich) solution and following the manufacturer's procedure. Samples scored positive were two-fold serially diluted to determine the endpoint titre.

4.3. Statistical Analysis

Statistical evaluation was carried out by the χ^2 test to analyze the results of serological tests in relationship to the years in which samples were collected. Values of $P < 0.05$ were considered significant.

5. Conclusions

Grazing horses living in central Italy seem to be frequently exposed to ticks, and consequently to pathogens transmitted by these hematophagous vectors. The present results confirm the spreading of *A. phagocytophilum* in the investigated geographic area, where the bacterium has been previously found in various animal populations.

Moreover, horses scored positive to SFG *Rickettsia* spp. suggest that they can be infected by this microorganism, too, even though they do not develop disease.

In all cases, infected horses may be involved in the epidemiological cycle of *A. phagcoytophilum* and SFG species, as *R. conorii* and other rickettsiae that were considered non-pathogenic for decades and now are associated with human infections.

Considering that the spreading of tick-borne diseases is a growing concern, and in light of the One-Health concept, it is necessary to increase the surveillance of these infections in animals, not only pets but also horses, that have frequent contact with humans.

Funding: This research received no external funding.

Conflicts of Interest: The author declares no conflict of interest.

References

1. Dumler, J.S.; Barbet, A.F.; Bekker, C.P.J.; Dasch, G.A.; Palmer, G.H.; Ray, S.C.; Rikihisa, Y.; Rurangirwa, F.R. Reorganization of genera in the families *Rickettsiaceae* and *Anaplasmataceae* in the order *Rickettsiales*: Unification of some species of *Ehrlichia* with *Anaplasma*, *Cowdria* with *Ehrlichia*, and *Ehrlichia* with *Neorickettsia*, descriptions of six new species combinations and designation of *Ehrlichia equi* and "HGE agent" as subjective synonyms of *Ehrlichia phagocytophila*. *Int. J. Syst. Evol. Microbiol.* **2001**, *51*, 2145–2165.
2. Stuen, S.; Granquist, E.G.; Silaghi, C. *Anaplasma phagocytophilum*—A wide spread multi-host pathogen with highly adaptive strategies. *Front. Cell. Infect. Microbiol.* **2013**, *3*, 1–33. [CrossRef]
3. Bakken, J.B.; Dumler, J.S. Human granulocytic anaplasmosis. *Infect. Dis. Clin. N. Am.* **2015**, *29*, 341–355. [CrossRef]
4. Franzén, P.; Aspan, A.; Egenvall, A.; Gunnarsson, A.; Aberg, L.; Pringle, J. Acute clinical, hematologic, serologic, and polymerase chain reaction findings in horses experimentally infected with a European strain of *Anaplasma phagocytophilum*. *J. Vet. Intern. Med.* **2005**, *9*, 232–239. [CrossRef]
5. Brouqui, P.; Parola, P.; Fournier, P.E.; Raoult, D. Spotted fever rickettsioses in southern and Eastern Europe. *FEMS Immunol. Med. Microbiol.* **2007**, *49*, 2–12. [CrossRef]
6. Vescio, M.F.; Piras, M.A.; Ciccozzi, M.; Carai, A.; Farchi, F.; Maroli, M.; Mura, M.S.; Rezza, G.; MSF Study Group. Socio-demographic and climatic factors as correlates of Mediterranean spotted fever (MSF) in northern Sardinia. *Am. J. Trop. Med. Hyg.* **2008**, *78*, 318–320. [CrossRef]
7. Beninati, T.; Lo, N.; Noda, H.; Esposito, F.; Rizzoli, A.; Favia, G.; Genchi, C. First detection of spotted fever group rickettsiae in *Ixodes ricinus* from Italy. *Emerg. Infect. Dis.* **2002**, *8*, 983–986. [CrossRef]
8. Vitale, G.; Mansuelo, S.; Rolain, J.M.; Raoult, D. *Rickettsia massiliae* human isolation. *Emerg. Infect. Dis.* **2006**, *12*, 174–175. [CrossRef]
9. Floris, R.; Yurtman, A.N.; Margoni, E.F.; Mignozzi, K.; Boemo, B.; Altobelli, A.; Cinco, M. Detection and identification of *Rickettsia* species in the northeast of Italy. *Vector Borne Zoonotic Dis.* **2008**, *8*, 777–782. [CrossRef]
10. Mura, A.; Masala, G.; Tola, S.; Satta, G.; Fois, F.; Piras, P.; Rolain, J.M.; Raoult, D.; Parola, P. First direct detection of rickettsial pathogens and a new rickettsia, "Candidatus *Rickettsia barbariae*", in ticks from Sardinia, Italy. *Clin. Microbiol. Infect.* **2008**, *14*, 1028–1033.
11. Selmi, M.; Martello, E.; Bertolotti, L.; Bisanzio, D.; Tomassone, L. *Rickettsia slovaca* and *Rickettsia raoultii* in *Dermacentor marginatus* ticks collected on wild boars in Tuscany, Italy. *J. Med. Entomol.* **2009**, *46*, 1490–1493. [CrossRef]
12. Trotta, M.; Nicetto, M.; Fogliazza, A.; Montarsi, F.; Caldin, M.; Furlanello, T.; Solano-Gallego, L. Detection of *Leishmania infantum*, *Babesia canis*, and rickettsiae in ticks removed from dogs living in Italy. *Ticks Tick Borne Dis.* **2012**, *3*, 294–297. [CrossRef]
13. Elfving, K.; Malmsten, J.; Dalin, A.M.; Nilsson, K. Serological and molecular prevalence of *Rickettsia helvetica* and *Anaplasma phagocytophilum* in wild cervids and domestic mammals in the Central parts of Sweden. *Vector Borne Zoonotic Dis.* **2015**, *15*, 529–534. [CrossRef]
14. Skotarczak, B.; Wodecka, B.; Rymaszewska, A.; Adamska, M. Molecular evidence for bacterial pathogens in *Ixodes ricinus* ticks infesting Shetland ponies. *Exp. Appl. Acarol.* **2016**, *69*, 179–189. [CrossRef]
15. Parola, P.; Paddock, C.D.; Socolovschi, C.; Labruna, M.B.; Mediannikov, O.; Kernif, T.; Abdad, M.Y.; Stenos, J.; Bitam, I.; Fournier, P.E.; et al. Update on tick-borne rickettsioses around the world: A geographic approach. *Clin. Microbiol. Rev.* **2013**, *26*, 657–702, Erratum in: *Clin. Microbiol. Rev.* **2014**, *27*, 166. [CrossRef]

16. Alberti, A.; Zobba, R.; Chessa, B.; Addis, M.F.; Sparagano, O.; Pinna Parpaglia, M.L.; Cubeddu, T.; Pintori GPittau, M. Equine and canine *Anaplasma phagocytophilum* strains isolated on the island of Sardinia (Italy) are phylogenetically related to pathogenic strains from the United States. *Appl. Environ. Microbiol.* **2005**, *71*, 6418–6422. [CrossRef]

17. Passamonti, F.; Veronesi, F.; Cappelli, K.; Capomaccio, S.; Coppola, G.; Marenzoni, M.L.; Piergili Fioretti, D.; Verini Supplizi, A.; Coletti, M. *Anaplasma phagocytophilum* in horses and tick: A preliminary survey of central Italy. *Comp. Immunol. Microbiol. Infect. Dis.* **2010**, *33*, 73–83. [CrossRef]

18. Giudice, E.; Giannetto, C.; Furco, V.; Alongi, A.; Torina, A. *Anaplasma phagocytophilum* seroprevalence in equids: A survey in Sicily (Italy). *Parasitol. Res.* **2012**, *111*, 951–955. [CrossRef]

19. Laus, F.; Veronesi, F.; Passamonti, F.; Paggi, E.; Cerquetella, M.; Hyatt, D.; Tesei, B.; Fioretti, D.P. Prevalence of tick borne pathogens in horses from Italy. *J. Vet. Med. Sci.* **2013**, *75*, 715–720. [CrossRef]

20. Ebani, V.V.; Nardoni, S.; Bertelloni, F.; Rocchigiani, G.; Mancianti, F. Tick-borne infections in horses from Tuscany, Italy. *J. Equine Vet. Sci.* **2015**, *35*, 290–294. [CrossRef]

21. Leblond, A.; Pradier, S.; Pitel, P.H.; Fortier, G.; Boireau, P.; Chadoeuf, J.; Sabatier, P. An epidemiological survey of equine anaplasmosis (*Anaplasma phagocytophilum*) in southern France. *Rev. Sci. Tech.* **2005**, *24*, 899–908. [CrossRef]

22. Ribeiro, A.J.; Cardoso, L.; Maia, J.M.; Coutinho, T.; Cotovio, M. Prevalence of *Theileria equi, Babesia caballi* and *Anaplasma phagocytophilum* in horses from the north of Portugal. *Parasitol. Res.* **2013**, *112*, 2611. [CrossRef]

23. Tsachev, I.; Pantchev, N.; Marutsov, P.; Petrov, V.; Gundasheva, D.; Baymakova, M. Serological evidence of *Borrelia burgdorferi, Anaplasma phagocytophilum* and *Ehrlichia* spp. infections on horses from Southeastern Bulgaria. *Vector Borne Zoonotic Dis.* **2018**, *18*, 588–594. [CrossRef]

24. Hulínská, D.; Langøová, K.; Pejèoch, M.; Pavlásek, I. Detection of *Anaplasma phagocytophilum* in animals by real-time polymerase chain reaction. *APMIS* **2004**, *112*, 239–247. [CrossRef]

25. Butler, C.M.; Nijhof, A.M.; Jongejan, F.; Van der Kolk, J.H. *Anaplasma phagocytophilum* infection in horses in the Netherlands. *Vet. Rec.* **2008**, *162*, 216–217. [CrossRef]

26. Slivinska, K.; Vichová, B.; Wersko, J.; Szewczyk, T.; Wróblewski, Z.; Pet'ko, B.; Ragač, O.; Demeshkant, V.; Karbowiak, G. Molecular surveillance of *Theileria equi* and *Anaplasma phagocytophilum* infections in horses from Ukraine, Poland and Slovakia. *Vet. Parasitol.* **2016**, *215*, 35–37. [CrossRef]

27. Rosso, F.; Tagliapietra, V.; Baráková, I.; Derdáková, M.; Konečny, A.; Hauffe, H.C.; Rizzoli, A. Prevalence and genetic variability of *Anaplasma phagocytophilum* in wild rodents from the Italian Alps. *Parasites Vectors* **2017**, *10*, 293. [CrossRef]

28. De Arcangeli, S.; Balboni, A.; Serafini, F.; Battilani, M.; Dondi, F. *Anaplasma phagocytophilum* infection in thrombocytopenic dogs. *Vet. Ital.* **2018**, *54*, 73–78.

29. Ebani, V.V.; Rocchigiani, G.; Bertelloni, F.; Nardoni, S.; Leoni, A.; Nicoloso, S.; Mancianti, F. Molecular survey on the presence of zoonotic arthropod-borne pathogens in wild red deer (*Cervus elaphus*). *Comp. Immunol. Microbiol.* **2016**, *47*, 77–80. [CrossRef]

30. Ebani, V.V.; Rocchigiani, G.; Nardoni, S.; Bertelloni, F.; Vasta, V.; Papini, R.A.; Verin, R.; Poli, A.; Mancianti, F. Molecular detection of tick-borne pathogens in wild red foxes (*Vulpes vulpes*) from Central Italy. *Acta Trop.* **2017**, *172*, 197–200. [CrossRef]

31. Aureli, S.; Foley, J.E.; Galuppi, R.; Rejmanek, D.; Bonoli, C.; Tampieri, M.P. *Anaplasma phagocytophilum* in ticks from parks in the Emilia-Romagna region of northern Italy. *Vet. Ital.* **2012**, *48*, 413–423. [PubMed]

32. Pusterla, N.; Madigan, J.E. Anaplasma Phagocytophila. In *Equine Infectious Disease*; Sellon, C.D., Long, M.T., Eds.; Saunders: St Louis, MO, USA, 2007; pp. 354–357.

33. Gomez-Barroso, D.; Vescio, M.F.; Bella, A.; Ciervo, A.; Busani, L.; Rizzo, C.; Rezza, G.; Pezzotti, P. Mediterranean spotted fever rickettsiosis in Italy, 2001-2015: Spatio-temporal distribution based on hospitalization records. *Ticks Tick-Borne Dis.* **2019**, *10*, 43–50. [CrossRef] [PubMed]

34. Solano-Gallego, L.; Trotta, M.; Caldin, M.; Furlanello, T. Molecular survey of *Rickettsia* spp. *in sick dogs in Italy. Zoonoses Public Health* **2008**, *55*, 521–525. [PubMed]

35. Solano-Gallego, L.; Caprì, A.; Pennisi, M.G.; Caldin, M.; Furlanello, T.; Trotta, M. Acute febrile illness is associated with *Rickettsia* spp. infection in dogs. *Parasites Vectors* **2015**, *8*, 216. [CrossRef] [PubMed]

36. Riveros-Pinilla, D.A.; Acevedo, G.L.; Londoño, A.F.; Gongóra, A. Antibodies against spotted fever group *Rickettsia* sp. in horses of the Colombian Orinoquia. *Rev. MVZ Cordoba* **2015**, *20*, 5004–5013. [CrossRef]

37. Souza, C.E.; Bonato Camargo, L.; Pinter, A.; Donalisio, M.R. High seroprevalence for *Rickettsia rickettsii* in equine suggests risk of human infection in silent areas for the Brazilian Spotted Fever. *PLoS ONE* **2016**, *11*, e0153303. [CrossRef] [PubMed]

38. Filho, E.F.A.; Costa, F.B.; Moraes-Filho, J.; Gomes dos Santos, A.C.; Lopes do Vale, T.; Pereira da Costa, A.; Barbosa Silva, A.; Labruna, M.B.; de Maria Seabra Nogueira, R. Exposure of Baixadeiro horses ro Rickettsia spp. and to ticks infected by Rickettsia amblyommatis in the Baixada Maranhense micro-region, Maranhão, Brazil. *Cienc. Rural* **2018**, *48*, 1–7.

39. Ueno, T.E.H.; Costa, F.B.; Moraes-Filho, J.; Agostinho, W.C.; Fernandes, W.R.; Labruna, M.B. Experimental infection of horses with Rickettsia rickettsia. *Parasites Vectors* **2016**, *9*, 499. [CrossRef] [PubMed]

40. Vitaliti, G.; Falsaperla, R.; Lubrano, R.; Rapisarda, V.; Cocuzza, S.; Nunnari, G.; Pavone, P. Incidence of Mediterranean spotted fever in Sicilian children: A clinical-epidemiological observational retrospective study from 1987 to 2010. *Int. J. Infect. Dis.* **2015**, *31*, 35–40. [CrossRef] [PubMed]

41. Vascellari, M.; Ravagnan, S.; Carminato, A.; Cazzin, S.; Carli, E.; Da Rold, G.; Lucchese, L.; Natale, A.; Otranto, D.; Capelli, G. Exposure to vector-borne pathogens in candidate blood donor and free-roaming dogs of northeast Italy. *Parasites Vectors* **2016**, *9*, 369. [CrossRef]

42. La Scola, B.; Raoult, D. Laboratory diagnosis of rickettsioses: Current approaches to the diagnosis of old and new rickettsial diseases. *J. Clin. Microbiol.* **1997**, *35*, 2715–2727. [PubMed]

43. Lakos, A.; Kőrösi, A.; Földvári, G. Contact with horses is a risk factor for tick-borne lymphadenopathy (TIBOLA): A case control study. *Wien. Klin. Wochenschr.* **2012**, *124*, 611–617. [CrossRef] [PubMed]

44. Franzen, P.; Berg, A.L.; Aspan, A.; Gunnarsson, A.; Pringle, J. Death of a horse infected experimentally with *Anaplasma phagocytophilum*. *Vet. Rec.* **2007**, *160*, 122–125. [CrossRef] [PubMed]

 *pathogens*

Article

Occurrence of Bovine Cysticercosis in Two Regions of the State of Tocantins-Brazil and the Importance of Pathogen Identification

Benta Natânia Silva FIGUEIREDO [1,†], Ricardo Alencar LIBóRIO [2,†], Megumi SATO [3], Camila Figueira da SILVA [1], Ronaldo Alves PEREIRA-JUNIOR [1,4], Yuichi CHIGUSA [5], Satoru KAWAI [5] and Marcello Otake SATO [5,*,†]

1 Escola de Medicina Veterinária e Zootecnia, Universidade Federal do Tocantins, Araguaína 77804-970, Tocantins, Brazil; benta_naty@hotmail.com (B.N.S.F.); camila21_fi@hotmail.com (C.F.d.S.); ronaldo_pgtu@hotmail.com (R.A.P.-J.)
2 Instituto Federal do Tocantins, Gurupi 77410-470, Tocantins, Brazil; ricardolib_63@hotmail.com
3 Graduate School of Health Sciences, Niigata University, Niigata 951-8518, Niigata, Japan; satomeg@clg.niigata-u.ac.jp
4 Centro Universitário de Goiás, Uni-Anhanguera, Goiânia 74423-115, Goiás, Brazil
5 Department of Tropical Medicine and Parasitology, Dokkyo Medical University, Mibu 321-0293, Tochigi, Japan; ychigusa@dokkyomed.ac.jp (Y.C.); skawai@dokkyomed.ac.jp (S.K.)
* Correspondence: marcello@dokkyomed.ac.jp; Tel.: +81-282-872134; Fax: +81-282-86-6431
† These authors contributed equally to this work.

Received: 11 April 2019; Accepted: 16 May 2019; Published: 20 May 2019

Abstract: Bovine cysticercosis, caused by *Taenia saginata* metacestodes, is the cause of significant economic losses to the meat production chain by condemnation and downgrading of infected carcasses. It is also a public health issue causing human taeniasis. This study evaluated the occurrence of bovine cysticercosis at the meat inspection procedures in slaughterhouses of south and north regions of the Tocantins State in Brazil. Specimens identified as cysts of *T. saginata* were collected and analyzed by molecular (PCR) and histopathological techniques. The cysts were collected from March to December of 2010 in slaughterhouses located in the cities of Alvorada (South) and Araguaína (North). The frequency of cystic lesions during the study was 0.033% (53/164,091) with 69.81% of calcified lesions and 30.9% of live cysts at meat inspection. From 14 samples submitted to molecular analysis, 28.57% (4/14) were positive for *T. saginata*. The histopathological analysis of the non-*T. saginata* samples showed lesions suggestive of granuloma and hydatid disease. The results indicated that the identification of the etiological agent is difficult by macroscopic inspection, emphasizing the need to associate specific diagnostic methods at meat inspection in abattoirs. In addition, species-specific PCR would be an effective tool for diagnosis, monitoring, and identifying cysticercosis, assisting the conventional tests.

Keywords: public health; *Cysticercus bovis*; *Taenia saginata*; Tocantins; food security

---

1. Introduction

Brazil has the second largest cattle herd in the world with about 209 million heads, the second largest producer and exporter of beef [1]. The state of Tocantins, located in the eastern Amazon region of Brazil (Figure 1A), has a cattle herd of 6.3 million heads, bred by extensive farming in wide areas of savanna, fed mainly in natural grass spread across all the regions of the state [2]. The quality of the meat produced in the Tocantins made the state a main producer of beef for Brazilian and international markets, reaching more than 20 Countries in Europe and Asia [3].

Bovine cysticercosis caused by *Taenia saginata* is an important cattle disease, causing economic losses to the meat production chain due to the condemnation and downgrading of infected carcasses [4]. Furthermore, this cestode causing taeniasis in humans is a health issue, especially in regions with poor sanitary conditions, associated with socioeconomic and cultural aspects as the consumption of undercooked or raw beef [5,6]. Humans are infected by ingestion of viable *T. saginata* cysts in meat, the adult worm develops in the small intestine and releases gravid proglottids full of eggs in the feces contaminating the environment in areas with no sewage treatment. Bovines are infected by ingestion of grass with eggs developing cysts in striated muscles, also called *Cysticercus bovis* [7].

Routinely, the diagnosis is made by macroscopic examination during the post-mortem inspection of carcasses; however, the method has been criticized by low sensitivity in detecting [8] and limited diagnostic capability [9]. On the other hand, molecular techniques such as PCR have high sensitivity and specificity, allowing effective identification and differentiation of species of *Taenia*, overcoming many disadvantages of conventional methods [10–12]. There are few studies on bovine cysticercosis in the state of Tocantins with different prevalence estimations varying from 0.02% to 10.23% [13,14]

In this sense, this study aimed to verify the occurrence of bovine cysticercosis in slaughterhouses and confirm how molecular identification could better identify cases of the disease in meat inspection, using cysts collected in the south and north regions of the State of Tocantins (Brazil) (Figure 1B).

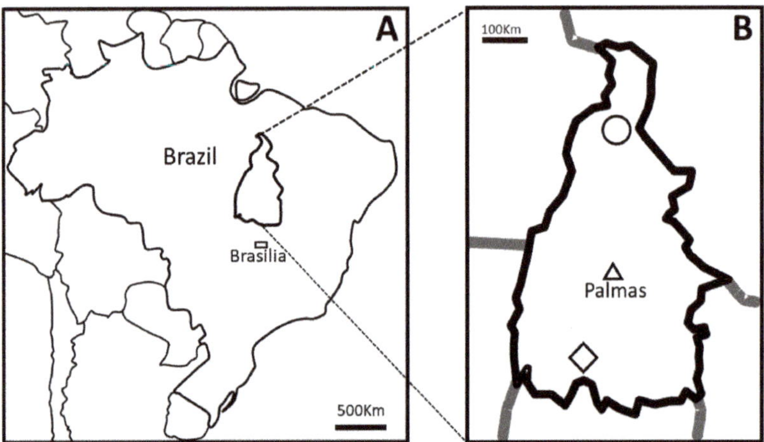

Figure 1. Map of Brazil indicating the position of the State of Tocantins (**A**) and the Country's capital city Brasília (rectangle). The material used in this study were collected in slaughterhouses in the municipalities of Araguaína (circle) in the north and Alvorada (diamond) in the south of Tocantins (**B**). The triangle in B indicates Palmas city, the capital of the State of Tocantins.

2. Results

In Table 1, the monthly number of bovines processed in each facility and the number of cases of viable and calcified cysticercosis referring to the period is shown. The total frequency of the bovine cysticercosis during the collection period (from March to December 2010) was 0.033% (53/164,091) with 30.19% (16/53) of live cysts and 69.81% (37/53) of calcified cysts. The prevalence of the bovine cysticercosis was 0.037% in Alvorada, with 0.015% live cysts (09 cases), and 0.022% calcified cysts (13 cases). The prevalence of live and calcified cysts was 0.006% (7 cases) and 0.023% (24 cases), respectively, in Araguaína, with a prevalence of 0.029% (Table 1). Based on the PCR technique, 28.57% (4/14) of collected cysts were identified as *T. saginata* (Figure 2, Table 2). In the present work, 90% (9/10) of the calcified cysts found at meat inspection were not detected by specific PCR.

Table 1. Total number of heads processed in the slaughterhouses of this study and the cases of viable (VI) and calcified (CA) cysticercosis diagnosed between March and September of 2010 by the inspection service.

Month	Alvorada	VI	CA	Araguaina	VI	CA
March	8564	01	01	15,788	03	01
April	8811	01	02	14,863	-	04
May	8772	02	-	13,379	-	01
June	9270	01	01	16,050	-	06
July	7731	02	04	18,428	02	07
August	7998	01	03	13,754	01	03
September	7445	01	02	13,238	01	02
Total	**58,591**	**09**	**13**	**105,500**	**07**	**24**

Table 2. Cyst samples collected for molecular analysis (PCR) from March to September 2010 in Alvorada city (A), Southern area, and Araguaina city (B) in the Northern area of the State of Tocantins, Brazil. The cysts were classified in viable (VI) and calcified (CA) and by the localization of the cyst.

Sample ID	Classification	Slaughterhouse	Collection Date	Local	PCR
C1	CA	B	March	Heart	P
C2	CA	B	April	Heart	N
C3	CA	B	April	Heart	N
C4	VI	A	April	Masseter	P
C5a	VI	A	July	Masseter	P
C5b	VI	A	July	Masseter	P
1a	CA	B	June	Masseter	N
1b	CA	B	June	Masseter	N
1c	CA	B	June	Masseter	N
2	VI	B	August	Heart	N
3	CA	A	August	Heart	N
4	CA	A	August	Masseter	N
5	CA	A	August	Masseter	N
6	CA	A	September	Liver	N

P–positive; N–negative.

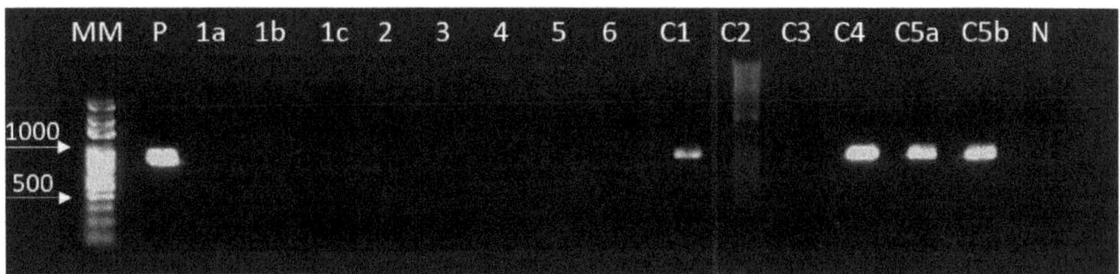

Figure 2. Amplification of *CO1* by *Taenia saginata* specific PCR in DNA purified from cystic lesions of bovines in Araguaina and Alvorada, Tocantins, Brazil. From left to right are; DNA size markers (MM), *Taenia saginata* DNA (accession no. AB107246) as a positive control (P), tested samples from 1a to c5b and bovine genomic DNA as a negative control (N). The cystic samples 1a, 1b, 1c, 2, 3, 4, 5, 6, C2, and C3 were not detectable using PCR in this study. The cysts C1, C4, C5a, and C5b were identified as *T. saginata* presenting an amplicon of 827 bp. C1 was the only calcified cyst presenting amplification by PCR.

The microscopic analysis of negative samples in the PCR for cysticercosis revealed that three samples showed lesions with characteristics of granuloma were observed multinuclear cells in a granulomatous lesion (Figure 3). In addition, three different cysts sections stained by PAS, showed laminated membranes and germinal layers in magenta color, a characteristic of hydatid cysts (Figure 4). Four samples presented inconclusive results.

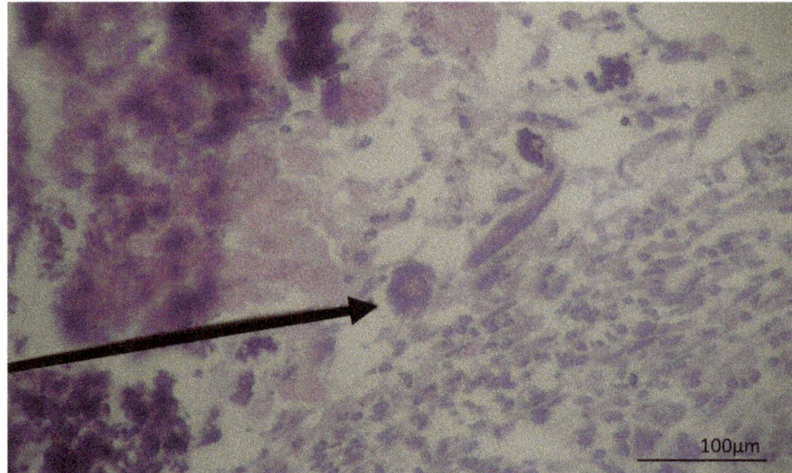

Figure 3. Photomicrography of a cyst recovered from a bovine liver showing a granulomatous lesion in hematoxylin and eosin stain. The arrow indicates Langhan's giant cells with broad cytoplasm and nuclei arranged at the periphery of the horseshoe-shaped cell.

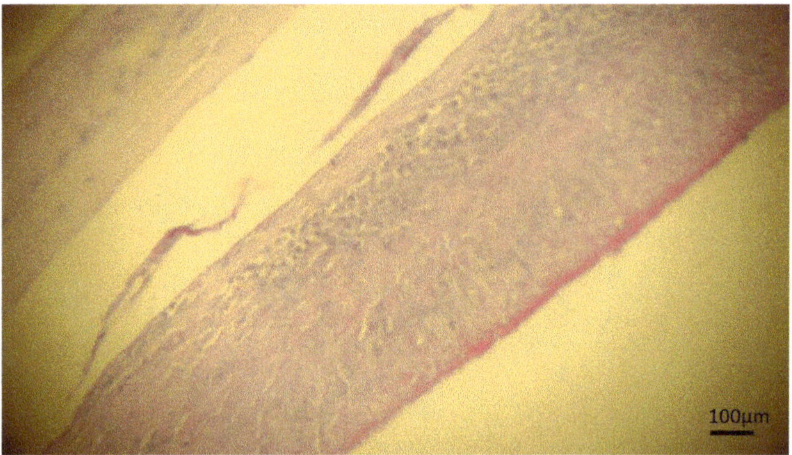

Figure 4. Photomicrography of cyst in bovine liver stained by periodic acid–Schiff stain (PAS). Continuous laminated membrane and fibrous capsule, a strong granulomatous reaction with the presence of germinative tissue is observed suggesting hydatidosis.

3. Discussion

The results found in this study showed a low frequency of bovine cysticercosis when compared with the data described by Marques et al. (2008) in a study with prevalence data in several Brazilian states, reporting a prevalence of 10.23 % of bovine cysticercosis in the State of Tocantins, which is higher than other Brazilian states as São Paulo (8.76%), Paraná (7.53%), Minas Gerais (5.92%), Mato Grosso do Sul (4.74%), Goiás (4.16%), and Mato Grosso (0.71%) [13]. However, the official data of the prevalence of cysticercosis cases in the state of the Tocantins in 2009 and 2010 was 0.02% [14], near our findings of 0.033% total prevalence. There was no targeted action directed for taeniasis/cysticercosis control in the state of Tocantins, and the sharp decrease of bovine cysticercosis prevalence in 10 years could be related with the improvements in education, health infrastructure built after the creation of the state in 1989 and the urbanization of the population from 40.1% in 1980 to 78.8% in 2010 [15].

The macroscopic examination of mineralized (calcified) cysts, generally classifying it as calcified cysticercosis may overestimate the results of the prevalence of the bovine cysticercosis, as a specific diagnosis of calcified cysts is very difficult by post-mortem inspection [16–18].

It is not possible to discard the possibility of cysticercosis indeed; however there is a higher possibility of false-positive results by eye inspection once macroscopically, calcified hepatic nodular lesions are difficult to diagnose due to intense inflammatory reaction [16]. In this aspect, the histopathological examination may help in the identification of the causative agent, with the visualization of microscopic characteristic structures, as a laminated membrane in hydatidosis and calcareous corpuscles in cysticercosis [16,18–20]. In calcified cysts, DNA can be detected (with a lower sensitivity) by PCR [19]. PCR could not detect most *T. saginata* in the calcified cysts in this study, probably because of the decrease in sensitivity by DNA degradation in old cysts.

The definitive diagnosis of hydatidosis is usually based on histopathological analysis using PAS staining, where the laminated membrane, germinal layers, and protoscolices are PAS-positive staining in magenta color [20,21], the membrane staining characteristics were observed in three cysts in the present study, however, protoscolices were not observed (Figure 4).

The presence of granuloma with multinucleated giant cells indicates chronic inflammation. In those granulomatous lesions, there is focal necrosis with mineralization, the presence of macrophages, epithelioid cells, and Langerhans cells resulting from the fusion of epithelioid cells, there are different origins for this type of pathological lesion including mycobacterium and other bacterial infections, fungi, and protozoa [22,23] but not cysticercosis. In our histopathological evaluation, despite not all characteristics of granuloma being observed, it is possible to detect multinucleated cells, suggesting the cystic lesion could be or evolve in a granuloma.

As demonstrated in the present study, the visual inspection may present limitations in the judgment of cysts, mainly those calcified, since these have similar macroscopic characteristics with various agents, this may cause distortions in data of occurrence and prevalence reports.

T. saginata was found in the two regions studied, in the North and South areas of the State of Tocantins. Although the low prevalence of bovine cysticercosis, the absolute numbers are relevant, indicating environmental contamination and the need for basic sanitary care. Twenty-two infected animals in the south region and 31 in the north region presented cysticercosis. These results indicate the presence of human taeniasis and the contamination of the environment with parasite eggs is probably due to the lack of sanitary education and inadequate waste treatment [24–26].

This study highlights the importance of associated diagnostic methods for improved diagnosis in meat inspection with molecular detection as an accurate alternative for monitoring data, which may be overestimated by conventional techniques.

4. Materials and Methods

4.1. Collection of Cysts

To evaluate the specificity of gross examination, fourteen samples of cystic lesions of cattle were collected during the post-mortem inspection in the slaughterhouses of Alvorada and Araguaína, which are the main processors of livestock from cities located in the south and north regions of the State of Tocantins (Brazil), respectively, from March to September of 2010 (Figure 1).

The post-mortem examination of the carcasses followed the routine protocols to cysticercosis investigation in the inspection line in abattoirs and constituted by the slicing and inspection of the muscles of head, tongue, heart, diaphragm, and esophagus, according to the regulations of the Department of Inspection of Animal Origin Products (DIPOA) [27].

The cysts were classified in live or calcified by morphologic characteristics at the inspection line and were dissected, packed, labeled, and stored at 4 °C before sending to the laboratory (Laboratório de Parasitologia Veterinária) of the Universidade Federal do Tocantins for further analysis.

4.2. DNA Extraction and PCR

Cysts were individually cut in small pieces, transferred to microtubes, and disrupted using a pestle. The tubes were incubated at 55 °C in lysis buffer (100mM Tris HCl pH 8.5, 0.5M EDTA, 10% SDS, 5M NaCl, 20mg/mL Proteinase K) until complete digestion. Genomic DNA of each cyst was extracted and purified using the phenol/chloroform/isoamyl alcohol method and purified by ethanol precipitation [28]. The identification of *T. saginata* was carried out using the species-specific PCR targeting the mitochondrial *CO1* with an amplicon of 827 bp as described previously [11] with some modification as follows: instead of multiplex PCR, the PCR was done using the specific forward primers to *T. saginata* (H05 TsagF) and the reverse primer (E03 RevCOI). PCR reactions were done using KOD Plus Neo (Toyobo, Japan), amplifications were conducted in 50 µl total volume using 1 µl of template DNA and 12.5 picomoles of each primer. The PCR conditions were: 1 min to 98 °C followed by 35 cycles of 94 °C 60 s, 58 °C 30 s, 72 °C 60 s and a final extension of 72 °C for 5 min. Controls of PCR were added in each run. *T. saginata* with Brazilian origin (AB107246) genomic DNA (1 ng) as a positive control [11] and bovine genomic DNA from our DNA panel (1 ng) as a negative control of the reactions. The amplicons were visualized in 1.5% agarose gels under UV light by ethidium bromide staining.

4.3. Histopathological Evaluation

Part of the cyst wall together with the adjacent tissue of liver and muscle were collected and fixed in 10% buffered formalin for histopathological studies. Tissue samples were prepared in paraffin and sections 5 µm thick were stained using hematoxylin and eosin (HE) and periodic acid-Schiff stain [29].

Author Contributions: Conceptualization, M.O.S., R.A.L., and B.N.S.F; methodology, M.O.S., B.N.S.F., and R.A.L.; software, B.N.S.F.; validation, B.N.S.F., R.A.P.-J., C.F.d.S., and M.O.S.; formal analysis, M.O.S., M.S., and B.N.S.F.; investigation, R.A.L., B.N.S.F., C.F.d.S., and R.A.P.-J.; resources, R.A.L.; data curation, M.O.S., B.N.S.F., and R.A.L.; writing—original draft preparation, R.A.L. and B.N.S.F.; writing—review and editing, Y.C., M.O.S., M.S., S.K.; visualization, B.N.S.F.; supervision, M.O.S., M.S., and Y.C.; project administration, M.O.S.; funding acquisition, M.O.S.

Funding: This study was supported by grants from CNPq (578447/2008-8 and 502001/2009-7) and CAPES (Code 001).

Conflicts of Interest: The authors declare no conflict of interest.

References

1. Associação Brasileira das Indústrias Exportadoras de Carnes (ABIEC). São Paulo. Available online: http://www.abiec.com.br/3_rebanho.asp (accessed on 5 February 2014).
2. Instituto Brasileiro de Geografia e Estatística (IBGE). Censo Agropecuario 2017. Available online: https://www.ibge.gov.br/estatisticas-novoportal/economicas/agricultura-e-pecuaria/21814-2017-censo-agropecuario.html (accessed on 5 November 2018).
3. *Secretaria da Agricultura, da Pecuária e do Desenvolvimento Agrário—Tocantins*; SEAGRO-TO: Palmas, Brazil, 2018; Available online: https://seagro.to.gov.br/pecuaria (accessed on 5 November 2018).
4. Dorny, P.; Praet, N. *Taenia saginata* in Europe. *Vet. Parasitol.* **2007**, *149*, 22–24. [CrossRef]
5. de Aragão, S.C.; Biondi, G.F.; Lima, L.G.F.; Nunes, C.M. Animal cysticercosis in indigenous Brazilian villages. *Rev. Bras. Parasitol. Vet.* **2010**, *19*, 132–134. [CrossRef]
6. Ferreira, M.M.; Revoredo, T.B.; Ragazzi, J.P.; Soares, V.E.; Ferraldo, A.S.; Mendonça, R.F.; Lopes, W.D.Z. Prevalência, distribuição espacial e fatores de risco para cisticercose bovina no estado de São Paulo. *Pesq. Vet. Bras.* **2014**, *34*, 1181–1185. [CrossRef]
7. Sato, M.O.; Nunes, C.M.; Sato, M. Waikagul JTaenia. In *Biology of Foodborne Parasites*; Xiao, L., Ryan, U., Feng, Y., Eds.; EUA. CRC Press Taylor & Francis Group: New York, NY, USA, 2015; pp. 463–480.

8. Geysen, D.; Kanobana, K.; Victor, B.; Rodriguez-Hidalgo, R.; De Borchgrave, J.; Brandt, J.; Dorny, P. Validation of Meat Inspection Results for *Taenia saginata Cysticercosis* by PCR–Restriction Fragment Length Polymorphism. *J. Food Prot.* **2007**, *70*, 236–240. [CrossRef] [PubMed]

9. Abuseir, S.; Epe, C.; Schnieder, T.; Klein, G.; Kühne, M. Visual diagnosis of *Taenia saginata* cysticercosis during meat inspection: Is it unequivocal? *Parasitol. Res.* **2006**, *99*, 405–409. [CrossRef] [PubMed]

10. González, L.M.; Estrella, M.; Morakote, N.; Puente, S.; De Tuesta, J.L.D.; Serra, T.; Lopez-Velez, R.; McManus, D.P.; Harrison, L.J.S.; Parkhouse, R.M.E.; et al. Differential diagnosis of *Taenia saginata* and *Taenia saginata asiatica* taeniasis through PCR. *Diagn. Microbiol. Infect. Dis.* **2004**, *49*, 183–188. [CrossRef] [PubMed]

11. Yamasaki, H.; Allan, J.C.; Sato, M.O.; Nakao, M.; Sako, Y.; Nakaya, K.; Qiu, D.; Mamuti, W.; Craig, P.S.; Ito, A. DNA differential diagnosis of taeniasis and cysticercosis by multiplex PCR. *J. Clin. Microbiol.* **2004**, *42*, 548–553. [CrossRef] [PubMed]

12. Sato, M.O.; Sato, M.; Yanagida, T.; Waikagul, J.; Pongvongsa, T.; Sako, Y.; Sanguankiat, S.; Yoonuan, T.; Kounnavang, S.; Kawai, S.; et al. *Taenia solium, Taenia saginata, Taenia asiatica*, their hybrids and other helminthic infections occurring in a neglected tropical diseases' highly endemic area in Lao PDR. *PLoS Negl. Trop. Dis.* **2018**, *12*, 1–14. [CrossRef] [PubMed]

13. Marques, G.M.; Buzi, K.A.; Galindo, L.A.; Baldini, E.D.; Biondi, G.F. Avaliação dos Registros de condenação por cisticercose em bovinos abatidos em frigoríficos da região centro oeste do estado de São Paulo—1996–2000. *Vet. Zootec.* **2008**, *15*, 114–120.

14. SIGSIF/SISA/DDA/SFA/TO/MAPA. *Sistema de Informações Gerenciais do Serviço de Inspeção Federal/Serviço de Inspeção e Saúde Animal/Departamento de Defesa Animal/Superintendência Federal de Agricultura do Estado do Tocantins/Ministério da Agricultura, Pecuária e Abastecimento*; MAPA: Brasília, Brazil, 2011. Available online: http://sigsif.agricultura.gov.br/primeira_pagina/extranet/SIGSIF.html (accessed on 5 February 2014).

15. Instituto Brasileiro de Geografia e Estatística (IBGE). *Sinopse do Censo Demográfico 2010*; Instituto Brasileiro de Geografia e Estatística: Rio de Janeiro, Brazil, 2011; p. 261.

16. Almeida, D.O.; Igreja, H.P.; Alves, F.M.X.; Santos, I.F.; Tortelly, R. Cisticercose bovina em matadouro-frigorífico sob inspeção sanitária no município de Teixeira de Freitas-BA: Prevalência da enfermidade e análise anatomopatológica de diagnósticos sugestivos de cisticercose. *Revtst. Bras. Cienc. Vet.* **2006**, *13*, 178–182. [CrossRef]

17. Costa, R.F.R.; Santos, I.F.; Nascimento, E.R.; Tortelly, R. Caracterização das reações inflamatórias em corações de bovinos comercializados em açougues na cidade de Nova Friburgo/RJ. *Revtst. Bras. Cienc. Vet.* **2006**, *13*, 76–79.

18. Costa, R.F.R.; Santos, I.F.; Santana, A.P.; Tortelly, R.; Nascimento, E.R.; Fukuda, R.T.; Carvalho, E.C.Q.; Menezes, R.C. Caracterização das lesões por *Cysticercus bovis*, na inspeção *post mortem* de bovinos, pelos exames macroscópico, histopatológico e pela reação em cadeia da polimerase (PCR). *Pesq. Vet. Bras.* **2012**, *32*, 477–484. [CrossRef]

19. Yamasaki, H.; Nagase, T.; Kiyoshige, Y.; Suzuki, M.; Nakaya, K.; Itoh, Y.; Sako, Y.; Nakao, M.; Ito, A. A case of intramuscular cysticercosis diagnosed definitively by mitochondrial DNA analysis of extremely calcified cysts. *Parasitol. Int.* **2006**, *55*, 127–130. [CrossRef] [PubMed]

20. Schneider, R.; Gollackner, B.; Edel, B.; Schmid, K.; Wrba, F.; Tucek, G.; Walochnik, J.; Auer, H. Development of a new PCR protocol for the detection of species and genotypes (strains) of *Echinococcus* in formalin-fixed, paraffin-embedded tissues. *Int. J. Parasitol.* **2008**, *38*, 1065–1071. [CrossRef] [PubMed]

21. Ibrahim, S.E.A.; Gameel, A.A. (Eds.) Pathological, Histochemical and Immunohistochemical Studies of Lungs and Livers of Cattle and Sheep Infected with Hydatid Disease. In Proceedings of the 5th Annual Conference—Agricultural and Veterinary Research, Khartoum, Sudan, 24–27 February 2014.

22. Wangoo, A.; Johnson, L.; Gough, J.; Ackbar, R.; Inglut, S.; Hicks, D.; Spencer, Y.; Hewinson, G.; Vordermeier, M. Advanced Granulomatous Lesions in Mycobacterium bovis-infected Cattle are Associated with Increased Expression of Type I Procollagen, gd (WC1C) T Cells and CD 68C Cells. *J. Comp. Pathol.* **2005**, *133*, 223–234. [CrossRef] [PubMed]

23. Kumar, S.N.; Prasad, T.S.; Narayan, P.A.; Muruganandhan, J. Granuloma with langhans giant cells: An overview. *J. Oral. Maxillofac. Pathol.* **2013**, *17*, 420–423. [CrossRef] [PubMed]

24. Garcia-Garcia, M.L.; Torres, M.A.; Correa, D.O.; Flisser, A.; Sosa-Lechuga, A.L.; Velasco, O.S.; Meza-Lucas, A.N.; Plancarte, A.G.; Avila, G.U.; Tapia, R.A.; et al. Prevalence and risk of cysticercosis and taeniasis in an urban population of soldiers and their relatives. *Am. J. Trop. Med. Hyg.* **1999**, *61*, 386–389. [CrossRef] [PubMed]
25. Rossi, G.A.M.; Hoppe, E.G.L.; Mathias, L.A.; Martins, A.N.C.V.; Mussi, L.A.; Prata, L.F. Bovine cysticercosis in slaughtered cattle as an indicator of Good Agricultural Practices (GAP) and epidemiological risk factors. *Prev. Vet. Med.* **2015**, *118*, 504–508. [CrossRef] [PubMed]
26. Flisser, A.; Sarti, E.; Lightowlers, M.; Schantz, P. Neurocysticercosis: Regional status, epidemiology, impact and control measures in the Americas. *Acta Trop.* **2003**, *87*, 43–51. [CrossRef]
27. Brasil 1971. *Padronização de Técnicas, Instalações e Equipamentos. Volume 1. Bovinos. Serviço de Inspeção de produtos de Origem Animal (DIPOA)*; Ministério da Agricultura Pecuária e Abastecimento (MAPA): Brasília, Brazil, 1971; pp. 78–95.
28. Sambrook, J.; Fritsch, E.F.; Maniatis, T. *Molecular Cloning: A Laboratory Manual*, 9th ed.; Cold Spring Harbor Laboratory Press: New York, NY, USA, 1989; p. 545.
29. Drury, R.A.B.; Willington, E.A. *Carleton's Histological Technique*, 5th ed.; Oxford University Press: London, UK, 1980.

Article

Molecular Analysis of Canine Filaria and Its *Wolbachia* Endosymbionts in Domestic Dogs Collected from Two Animal University Hospitals in Bangkok Metropolitan Region, Thailand

Hathaithip Satjawongvanit [1], Atchara Phumee [2,3], Sonthaya Tiawsirisup [4], Sivapong Sungpradit [5], Narisa Brownell [2], Padet Siriyasatien [2] and Kanok Preativatanyou [2,*]

[1] Medical Science Program, Faculty of Medicine, Chulalongkorn University, Bangkok 10330, Thailand
[2] Vector Biology and Vector Borne Disease Research Unit, Department of Parasitology, Faculty of Medicine, Chulalongkorn University, Bangkok 10330, Thailand
[3] Thai Red Cross Emerging Infectious Disease-Health Science Centre, World Health Organization Collaborating Centre for Research and Training on Viral Zoonoses, Chulalongkorn Hospital, Bangkok 10330, Thailand
[4] Veterinary Parasitology Unit, Department of Veterinary Pathology, Faculty of Veterinary Science, Chulalongkorn University, Bangkok 10330, Thailand
[5] Department of Pre-clinic and Applied Animal Science, Faculty of Veterinary Science, Mahidol University, Salaya, Nakhon Pathom 73170, Thailand
* Correspondence: Kanok.Pr@chula.ac.th; Tel.: +66-2256-4387

Received: 4 June 2019; Accepted: 26 July 2019; Published: 29 July 2019

Abstract: Canine filariasis is caused by several nematode species, such as *Dirofilaria immitis*, *Dirofilaria repens*, *Brugia pahangi*, *Brugia malayi*, and *Acanthocheilonema reconditum*. Zoonotic filariasis is one of the world's neglected tropical diseases. Since 2000, the World Health Organization (WHO) has promoted a global filarial eradication program to eliminate filariasis by 2020. Apart from vector control strategies, the infection control of reservoir hosts is necessary for more effective filariasis control. In addition, many studies have reported that *Wolbachia* is necessary for the development, reproduction, and survival of the filarial nematode. Consequently, the use of antibiotics to kill *Wolbachia* in nematodes has now become an alternative strategy to control filariasis. Previously, a case of subconjunctival dirofilariasis caused by *Dirofilaria* spp. has been reported in a woman who resides in the center of Bangkok, Thailand. Therefore, our study aimed to principally demonstrate the presence of filarial nematodes and *Wolbachia* bacteria in blood collected from domestic dogs from the Bangkok Metropolitan Region, Thailand. A total of 57 blood samples from dogs with suspected dirofilariasis who had visited veterinary clinics in Bangkok were collected. The investigations for the presence of microfilaria were carried out by using both microscopic and molecular examinations. PCR was used as the molecular detection method for the filarial nematodes based on the *COI* and *ITS1* regions. The demonstration of *Wolbachia* was performed using PCR to amplify the *FtsZ* gene. All positive samples by PCR were then cloned and sequenced. The results showed that the filarial nematodes were detected in 16 samples (28.07%) using microscopic examinations. The molecular detection of filarial species using *COI*-PCR revealed that 50 samples (87.72%) were positive; these consisted of 33 (57.89%), 13 (22.81%), and 4 (7.02%) samples for *D. immitis*, *B. pahangi*, and *B. malayi*, respectively. While the *ITS1*-PCR showed that 41 samples (71.93%) were positive—30 samples (52.63%) were identified as containing *D. immitis* and 11 samples (19.30%) were identified to have *B. pahangi*, whereas *B. malayi* was not detected. Forty-seven samples (82.45%) were positive for *Wolbachia* DNA and the phylogenetic tree of all positive *Wolbachia* was classified into the supergroup C clade. This study has established fundamental data on filariasis associated with *Wolbachia* infection in domestic dogs in the Bangkok Metropolitan Region. An extensive survey of dog blood samples would provide valuable epidemiologic data on potential zoonotic filariasis in Thailand. In addition, this information could be used for the future development of more effective prevention and control strategies for canine filariasis in Thailand.

Keywords: dog; filariasis; *D. immitis*; *B. pahangi*; *B. malayi*; zoonosis; Thailand

1. Introduction

Filariasis is a mosquito-borne parasitic disease, which is an important public health concern in tropical and subtropical areas. It affects approximately 886 million people in 52 countries worldwide [1]. The disease is commonly found in Southeast Asian countries [2]. Filariasis has been reported as an important zoonosis from dogs and cats, both of which are well recognized as reservoirs for filariasis [3]. Several major canine filarial species have been documented in several localities around the world, including *Dirofilaria immitis*, *Dirofilaria repens*, *Brugia malayi*, *Brugia pahangi*, *Brugia ceylonensis*, *Brugia patei*, *Cercopithifilaria grassii*, *Acanthocheilonema reconditum*, and *Acanthocheilonema dracunculoides* [4–6]. In Thailand, the common filarial worms in dogs are *D. immitis*, *B. malayi*, *B. pahangi*, and *A. reconditum* [7–9]. *D. immitis* is the most common species; it causes canine heartworm disease or canine dirofilariasis and is widespread. The disease is endemic in tropical and subtropical regions throughout the world [10]. Among the neglected tropical diseases, filariasis was selected as a particular target for achieving elimination by 2020, and this has been published in the World Health Organization (WHO) roadmap [11,12]. Previously, the elimination and control strategy of filariasis was based on the control of the mosquito vectors, such as those in the *Mansonia*, *Anopheles*, *Aedes*, *Culex*, and *Ochlerotatus* genuses [13,14]. However, animals, especially dogs, can also serve as the reservoirs of filarial nematodes. Therefore, the control of infection in all potential reservoirs is also essential. Moreover, a study in dog reservoirs might provide a more in-depth understanding of the patterns and epidemiologic data of filariasis, and it could also contribute to a decrease in the number of human cases. Zoonotic filariasis has been reported globally. In Thailand, several human cases of zoonotic filariasis have been documented. Some examples include the ocular *Dirofilaria* infections caused by *Dirofilaria* spp. [15] and *D. repens* [16] in the patients from Phangnga and Nakhon Si Thammarat provinces, respectively. The most recently reported case was in 2018. A 67-year-old woman residing in Bangkok presented with subconjunctival dirofilariasis. Molecular analysis of the sample obtained from the eye of the patient showed that it was caused by an unknown *Dirofilaria* spp. closely related to *Dirofilaria hongkongensis* [17].

Several reports suggest that filarial nematodes serve as hosts for an obligate bacterial endosymbiont, *Wolbachia* [18–20]. *Wolbachia*, a gram-negative intracellular bacterium, plays an important role in nematode biological development, reproduction, and survival. It also provides critical metabolites to the filarial nematode [11,21,22]. In addition, the correlation between *Wolbachia*–nematode symbiosis has provided treatment strategies for the control and eradication of the filarial infection in the host using *Wolbachia* as a target of antibiotic treatment [23,24]. Many studies show that filarial nematode in dogs can be diagnosed through the examination of circulating microfilariae [25–27]. However, when the Giemsa stain is used for characterization, it is difficult to differentiate and identify definite species among the closely related microfilaria species because of the similarities in their morphology. As previously mentioned, there is a report on the zoonotic case in Bangkok caused by an unknown *Dirofilaria* species, and the morphology identification problem of the blood stage nematode and the relationship between nematodes and *Wolbachia* remain unsolved. Therefore, the aims of our study are primarily to determine the presence of filaria nematodes in domestic dog blood samples from the Bangkok Metropolitan Region, Thailand using molecular techniques to precisely identify the species of detected filarial nematodes. Polymerase chain reaction (PCR) was used as a molecular technique to amplify the cytochrome c oxidase subunit I (*COI*) gene and internal transcribed spacer 1 (*ITS1*) region, which are both suitable molecular markers given the high degree of genetic variation in filarial nematodes. The study was also used to investigate the status of *Wolbachia* symbionts in filarial nematodes in Thailand. The presence of *Wolbachia* bacteria was demonstrated by PCR to amplify the filamenting temperature-sensitive protein mutant Z (*FtsZ*) gene, which plays a role in the cell division of *Wolbachia* and can also be found in high copy numbers. Sequences obtained from the study were used for phylogenetic tree construction. Information obtained from the study provides preliminary data on filariasis associated with *Wolbachia* infection in domestic dogs in Bangkok, Thailand. These data will be useful in the future for reservoir- and *Wolbachia*-based programs for filariasis control.

2. Results

2.1. Morphological Characteristics of Filarial Worm

Microfilariae were detected in 16 of 57 (28.07%) dog blood samples collected in the Bangkok Metropolitan Region using the Giemsa staining technique. The morphological identification for microfilariae species found that 11 samples (19.30%) were *Dirofilaria* spp. and 5 samples (8.77%) were *Brugia* spp. (Table 1). For the *Dirofilaria* spp., Giemsa staining showed unsheathed microfilaria and the relative positions of the nerve ring (NR), excretory pore (Ex.P), excretory cell (Ex.C), first genital cell (G1), anal pore (AP), and the terminal nucleus (TN) at the tail. (Figure 1). For the *Brugia* spp., Giemsa staining showed clear sheathed microfilaria and one nucleus in the elongated cephalic space. The position of the nerve ring (NR), excretory pore (Ex.P), anal pore (AP), and two discrete overlapping terminal nuclei at the tail end are shown in Figure 1.

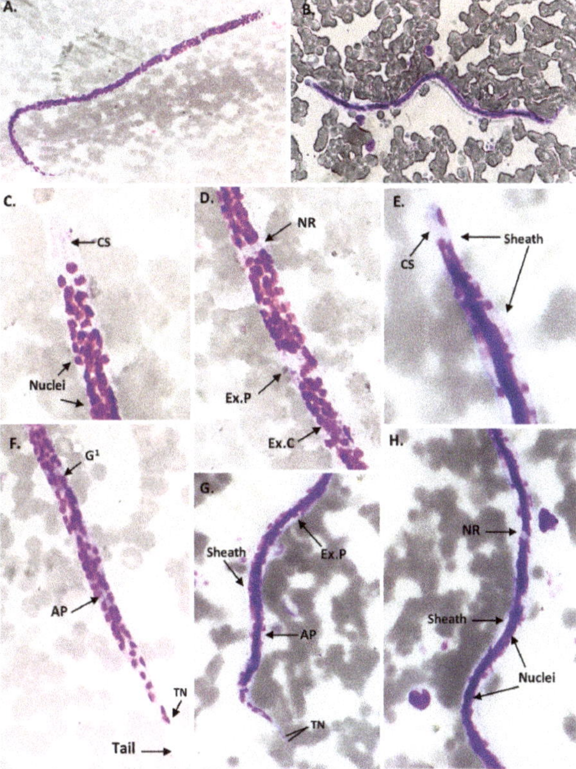

Figure 1. Microfilaria of *Dirofilaria* spp. (**A,C,D,F**) and *Brugia* spp. (**B,E,G,H**), Giemsa stain, 100×. The morphological marks show the position of several structures: The cephalic space (CS), nerve ring (NR), excretory pore (Ex.P), excretory cell (Ex.C), genital cell (G1), anal pore (AP), and terminal nucleus (TN).

2.2. Molecular Detection of Filarial Nematode

Both *COI*-PCR and *ITS1*-PCR were able to detect filarial nematode DNA in the blood samples. Of the 57 blood samples collected, 50 (87.72%) samples were amplified. The nucleotide sequences of the partial *COI* gene contained approximately 690 bp (Figure S1) and were able to classify filarial nematodes into three species: *D. immitis*, *B. pahangi*, and *B. malayi*, accounting for 33 (57.89%), 13 (22.81%), and 4 (7.02%) of the samples, respectively. Meanwhile, 41 samples (71.93%) were amplified for filarial nematode DNA using PCR annealed specifically to the *ITS1* region, which has approximately 600 bp for *D. immitis* (52.63%) and 550 bp for *B. pahangi* (19.30%) (Table 1, Figure S2). According to the sequence analysis of all sequences included in this study (Figures S3 and S4), the results of partial *COI* sequence comparisons between the species showed a sequence similarity percentage of 98.9%–100% (mean of 99.72%), 98.2%–100% (mean of 99.05%), and 98.6%–100% (mean of 99.22%) for *D. immitis*, *B. pahangi*, and *B. malayi* respectively, whereas the sequence similarity of *ITS1* was 83.3%–100% (mean of 92.2%) and 96.3%–100% (mean of 98.3%) for *D. immitis* and *B. pahangi*, respectively. The phylogenetic tree based on the partial *COI* and *ITS1* sequence showed that all isolated samples from *D. immitis* clustered together in one group. Two isolated from *B. pahangi* and *B. malayi* were clustered together, and they were clearly separated branches despite being placed in the same genus (Figures 2 and 3). The partial nucleotide sequences of the *COI* and *ITS1* obtained in this study were deposited in the GenBank under accession numbers MK250707–MK250757 and MK250758–MK250799, respectively. Based on the results, *COI*-PCR could be better suited for filarial species identification than *ITS1*-PCR. Additionally, present studies showed that *COI* evolves much more quickly than *ITS1*, revealing low intraspecific and interspecific molecular variability for filarial worms.

Table 1. The results of microscopic examinations and PCR for the detection of filarial worm DNA from blood dog specimens (n = 57). Positive controls used the DNA of *Dirofilaria immitis*, *Brugia pahangi*, and *Brugia malayi*.

Species	Microscopic Examination (%)	Molecular Detection (%)	
		COI	*ITS1*
Dirofilaria spp.	11(19.30%)		
Brugia spp.	5 (8.77%)		
D. immitis		33 (57.89%)	30 (52.63%)
B. pahangi		13 (22.81%)	11 (19.30%)
B. malayi		4 (7.02%)	0
Total	16 (28.07%)	50 (87.72%)	41 (71.93%)

2.3. Wolbachia Detection Using Nested PCR

Wolbachia DNA was detected based on the *Wolbachia* protein-coding housekeeping *FtsZ* gene (147 bp) (Figure S5). The result showed that 47 of 57 (82.45%) blood dog samples were positive for *Wolbachia* DNA. The species detected consisted of *D. immitis* (57.89%), *B. pahangi* (19.30%), and *B. malayi* (5.26%) (Table 2). The nucleotide sequence of *FtsZ* of the *Wolbachia* endosymbiont (Figure S6) had a similarity of 94.5%–100% (mean of 99.64%). The phylogenetic tree based on the *FtsZ* gene revealed that 47 of the *Wolbachia* DNA samples in this study could be clearly classified into supergroup C (Figure 4).

Table 2. Filarial worm tested for infection with *Wolbachia* using nested PCR of the *FtsZ* gene (n = 57). Positive controls used the DNA of *Wolbachia* supergroup B in *Aedes albopictus* mosquitoes.

Species	Positive for *Wolbachia* (%)
D. immitis	33 (57.89%)
B. pahangi	11 (19.30%)
B. malayi	3 (5.26%)
Total	47 (82.45%)

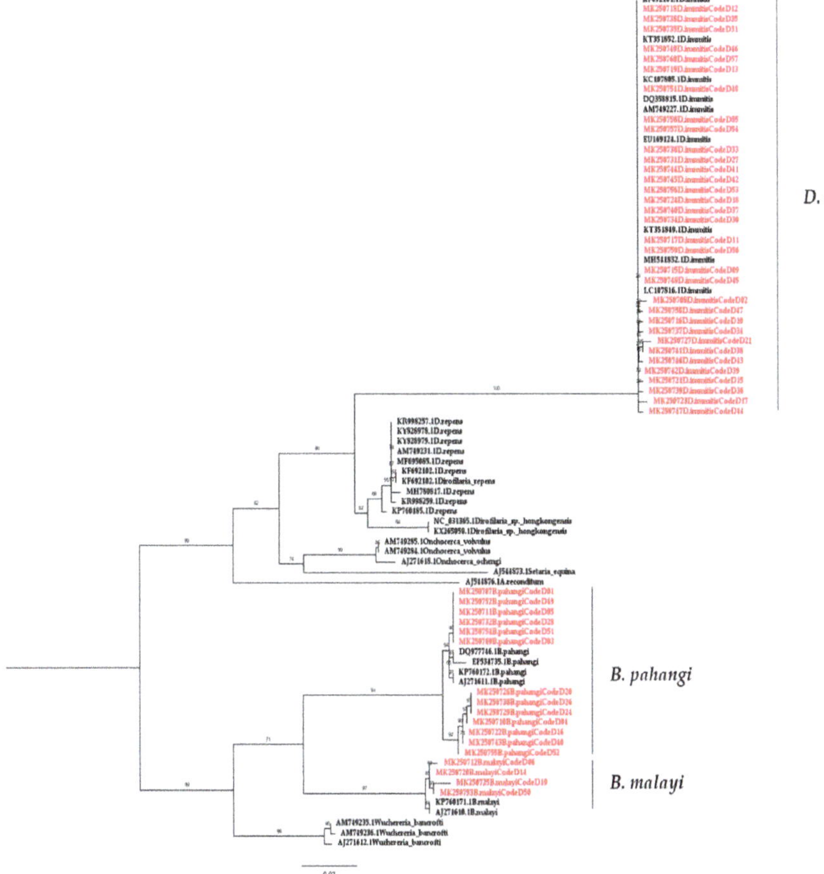

Figure 2. Phylogenetic tree of filarial worms constructed from the partial *COI* sequences. The red color represents the individual and combination identification sequences compared with reference isolates obtained from GenBank. Branch support was estimated based on 1000 bootstrap replicates.

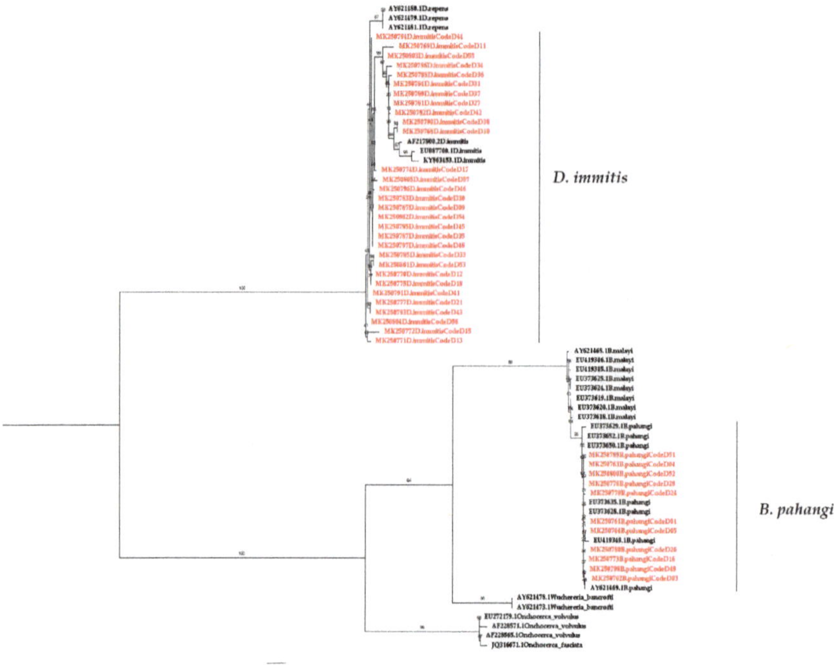

Figure 3. Phylogenetic tree of filarial worms constructed from *ITS1* sequences. The red color represents the individual and combination identification sequences compared with reference isolates obtained from GenBank. Branch support was estimated based on 1000 bootstrap replicates.

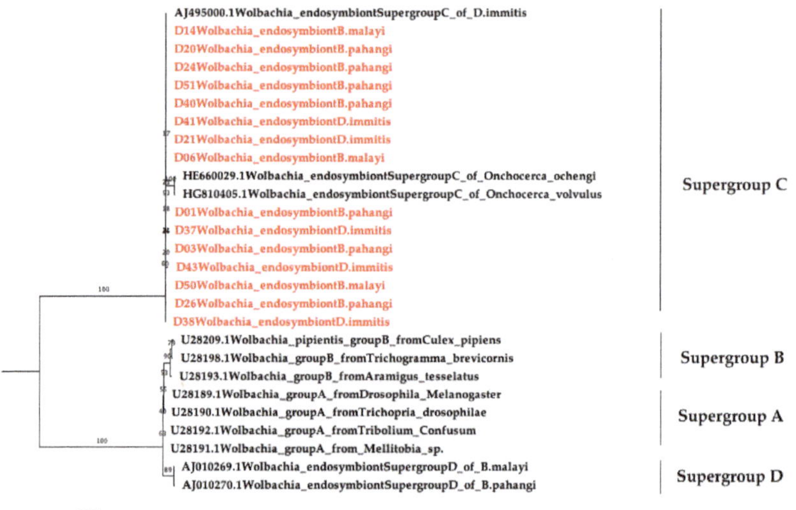

Figure 4. Phylogenetic tree of *Wolbachia* from filarial worms constructed from the partial *FtsZ* sequences. The red color represents the individual and combination identification sequences compared with reference isolates obtained from GenBank. Branch support was estimated based on 1000 bootstrap replicates.

3. Discussion

Canine filarioids are important nematodes that are transmitted to vertebrate hosts through arthropods and mosquitoes. Nematodes can cause severe disease in dogs and, potentially, transmit the disease to humans [28]. Control of canine filariasis through vector control and the early treatment of infected reservoir hosts is essential. Therefore, a highly specific and sensitive technique for the diagnosis of canine filarioids from an early stage of infection is necessary. Although the gold standard method for the diagnosis of microfilariae in a blood smear is microscopic examination on glass slides stained with Giemsa or hematoxylin and eosin [29,30], this technique requires considerable expertise, has low sensitivity, and cannot clearly discriminate among closely related species of filarial nematodes, such as *D. immitis, D. repens,* and *D. reconditum* or *B. malayi* and *B. pahangi* [31]. Serologic and molecular techniques with high sensitivity and specificity for filariasis diagnosis have been developed as additional methods for the detection of microfilariae [6,32–34]. In this study, microscopic examination gives a low positive result for microfilariae (28.07%) compared to PCR. We investigated conventional PCR for both canine filarial and *Wolbachia* bacterial infections in dog blood samples. PCR could be used to increase the detection rate of microfilaria DNA and also to differentiate filarial parasites among closely related species. This study also suggests that *COI* sequences can act as useful genetic markers in distinguishing between *D. immitis, B. pahangi,* and *B. malayi*; however, *ITS1* can detect only *D. immitis* and *B. pahangi*. Thus, the *COI*-PCR method is a more suitable method for the PCR-based detection of filarial nematode infections in dog blood specimens than *ITS1*-PCR. Therefore, we recommend using the *COI* gene for filarial detection because it is commonly used in the study of filarial DNA detection and has shown a large number of homolog sequences available in GenBank for the comparison and evaluation of results. Unfortunately, the limitation of this study was that the number of dog blood samples was quite low in number. To increase the accuracy of these assays, a larger sample size might be needed.

There are several reports on the prevalence of major *D. immitis* infection in dogs from various countries, including China [35], Korea [36], Iran [37], Portugal [38], Hungary [39], and Thailand [7–9,40]. In this current study, we also found that *D. immitis* is the most frequent species found in dog blood samples using both *COI*-PCR (57.89%) and *ITS1*-PCR (52.63%). This is followed by *B. pahangi*, detected in 22.81% of samples using *COI*-PCR and 19.30% of samples using *ITS1*-PCR. In contrast, *B. malayi* can be detected in only 7.02% of samples using *COI*-PCR. According to previous studies, both *B. malayi* and *B. pahangi* infections in dogs [41–43] and cats [7,9,44] have been found in different regions of southern Thailand, such as the Narathiwat and Satun provinces [7–9]. However, the information on filariasis in dogs caused by *B. malayi* and *B. pahangi* in terms of their relationship with their hosts, geographic distribution, and association with human or veterinary diseases is quite limited. Our study showed that the blood samples from dogs in the Bangkok Metropolitan Region were infected with both *B. malayi* and *B. pahangi*. Therefore, we suspect that domestic dogs may be a potential reservoir for *B. malayi* and *B. pahangi* in Bangkok. Further surveys throughout the country to investigate the prevalence of *B. malayi* and *B. pahangi* infection in a greater number of dog samples are needed in order to investigate the exact role of dogs as the natural reservoir hosts of *B. malayi* and *B. pahangi*. Further investigations on the mosquito vectors are also needed for the development of control protocols in the future.

The preliminary study of the phylogenetic tree based on the partial *COI* sequences of *B. malayi*–positive samples showed they were clustered together with *B. malayi* from France (accession no. KP760171.1) and Italy (accession no. AJ271610.1) with short branch lengths. The *B. pahangi* sequences of *COI* were also clustered together with *B. pahangi* from Malaysia (accession no. DQ977746.1 and EF534735.1), France (accession no. KP760172.1), and Italy (accession no. AJ271611.1) with small branch lengths. This is also true for *B. pahangi* sequences based on the *ITS1* region. According to the result of the phylogenetic tree of filarial worms constructed from partial *COI* sequences, two isolated samples from *B. malayi* and *B. pahangi* were in the same cluster and could be clearly separated. For this reason, the *COI* sequence was useful for identifying these congeneric species, and *COI*-PCR also showed better amplification for multiple species than *ITS1*-PCR in the present study. Oh et al. (2017) suggested that *COI* is the so-called "barcode" for the identification of filarial species diversity [45]. The strength of this study is that this method can be used as a screening and survey tool to study the molecular epidemiology of filarial infection among the high-risk population. In this study, we did not find *D. hongkongensis* as we expected. This might be due to the small number of samples, and other vertebrates may be the definitive host for *D. hongkongensis*. Interestingly, many filarial nematodes contain *Wolbachia*, an obligate bacterial endosymbiont, which plays an important role in the biological development of filarial nematodes and is involved in targeted therapy against them, leading to the loss of worm fertility and viability upon antibiotic treatment [23,24]. *Wolbachia* can be classified into at least seven supergroups based on the *FtsZ* and *Wolbachia* surface protein (*wsp*) sequences [46–49]. *Wolbachia* supergroups C and D have been detected in nematodes, whereas four supergroups (i.e., A, B, E, and H) have only been found in arthropods [50]. Furthermore, supergroup F has been found in both arthropod and nematode species [47]. Our results showed the preliminary study of the phylogenetic tree of all positive *Wolbachia* from 33 (57.89%), 11 (19.30%), and 3 (5.26%) samples of *D. immitis*, *B. pahangi*, and *B. malayi*, respectively. These were classified into a clade of supergroup C, which is closely related to the *Wolbachia* endosymbiont of *D. immitis* from Italy (accession no. AJ495000) and the *Wolbachia* endosymbiont of *Onchocerca ochengi* (accession no. HE660029) and *Onchocerca volvulus* (accession no. HG810405) from the United Kingdom [51]. In terms of *Wolbachia* and nematode symbiosis, it was revealed that *Wolbachia* are obligate mutualists for filariae worms and are necessary for the development, fertility, and vitality of adult filariae [52]. *Wolbachia* are found in all developmental stages of the filarial nematode but rapidly increase in number as the nematode transitions from its insect vectors to mammalian hosts. The *Wolbachia* titres further increase during the development of the larvae into the adult stages. The high titre of *Wolbachia* is also found within oocytes and infected embryos [53,54]. This study highlights PCR as the diagnostic tool for filariasis and the presence of *Wolbachia* in dogs from the Bangkok Metropolitan Region, Thailand. Regarding the limitations of the study, the examination of additional dog blood samples from geographical regions of Thailand is required for a more complete and robust understanding of the phylogeography of *Wolbachia* and filariasis in Thailand. The use of bacteria as a new control trend simultaneously targeting the vector and filarial parasites should also be explored in the future.

4. Materials and Methods

4.1. Ethics Statement

The study was approved by the animal research ethics committee of Chulalongkorn University Animal Care and Use Protocol (CU-ACUP), Faculty of Medicine, Chulalongkorn University, Bangkok, Thailand (COA No. 022/2561).

4.2. Sample Collection

A total of 57 blood samples were collected from domestic dogs in the Bangkok Metropolitan Region, Thailand by collaborating with veterinarians at the Small Animal Teaching Hospital, Chulalongkorn University and Faculty of Veterinary Science, Mahidol University Salaya Campus. The specimens were stored at 4 °C and transferred to the laboratory of the Vector Biology and Vector Borne Disease Research Unit, Department of Parasitology, Faculty of Medicine, Chulalongkorn University for filarial worm detection (Table S1). These samples were differentiated for morphological identification and molecular detection.

4.3. Microscopic Examination

Drops of blood were smeared on glass slides. Thick smears were fixed with absolute methanol (Sigma-Aldrich, St. Louis, MO, USA) for one minute, stained with Giemsa (Sigma-Aldrich, St. Louis, MO, USA) in a phosphate buffer (with a pH of 7.2), examined under a light microscope (CH20,

Olympus, Tokyo, Japan), and morphologically identified for microfilariae with or without sheath species using the taxonomic key [55–57].

4.4. DNA Extraction

DNA was extracted from the dog blood samples by using a blood DNA extraction kit (Invisorb Spin Blood Mini Kit, STRATEC Molecular, Berlin, Germany) according to the manufacturer's instructions. Extracted DNA was eluted in 50 µL of elution buffer. Extracted DNA samples were kept at −80 °C for long-term storage.

4.5. Molecular Identification of Canine Filarial

The microfilariae DNA samples were amplified for partial *COI* genes and *ITS1* regions. For the *COI* gene, two oligonucleotide primers from previous studies [58], COI-int-F (5′-TGATTGGTGGTTT TGGTAA-3′) and COI-int-R (5′-ATAAGTACGAGTATCAATATC-3′), were used for PCR amplification. The PCR reactions were performed in a total volume of 20 µL containing 50 ng of DNA template, 10× buffer (Thermo Fisher Scientific, Waltham, MA, USA), 25 mM of $MgCl_2$ (Thermo Fisher Scientific, Waltham, MA, USA), 2 mM of dNTPs (GeneAll, Korea), 10 µM each of forward and reverse primers, and five units of *Taq* DNA polymerase (Thermo Fisher Scientific, Waltham, MA, USA). The thermal profile was as follows: 94 °C for 3 min; 40 cycles of 94 °C for 45 s, 52 °C for 45 s, and 72 °C for 90 s; and 72 °C for 7 min. The *ITS1* region was PCR-amplified using forward primer ITS1-F (5′-GGTGAACCTGCGGAAGGATC-3′) and reverse primer ITS1-R (5′-CTCAATGCGTCTGCAATTCGC -3′), and this was performed in 25 µL of reaction mixture containing 50 ng of DNA template, 10× buffer (Thermo Fisher Scientific, Waltham, MA, USA), 25 mM of $MgCl_2$ (Thermo Fisher Scientific, MA, USA), 2 mM of dNTPs (GeneAll, Korea), 10 µM of each primer, and five units of *Taq* DNA polymerase (Thermo Fisher Scientific, Waltham, MA, USA). The PCR conditions were 94 °C for 5 min; 35 cycles of 94 °C for 30 s, 58 °C for 30 s, and 72 °C for 45 s; and 72 °C for 10 min [31]. The positive controls used in this study consisted of *D. immitis*, *B. pahangi*, and *B. malayi* DNA, whereas double-distilled water was used as a negative control. The PCR products were analyzed via 1.5% agarose gel electrophoresis, stained with ethidium bromide, and visualized with Quantity One Quantification Analysis Software Version 4.5.2 (Gel DocEQ System; Bio-Rad, Hercules, CA, USA).

4.6. Wolbachia Bacterial Detection

The nested PCR screening for *Wolbachia* bacteria was amplified by using two pairs of primer. Two primers, including Wol1-fwd (5′-CCTGTACTATATCCAAGAATTACTG-3′) and Wol1-R (5′-AC TATCCTTTATATGTTCCATAATTTC-3), were used for the first round. For the second round of amplification, two primers, Wol7-fwd (5′-GGTGGAAATGCTGTGAATAAC-3′) and Wol7-R (5′-AGC ACCGAGCCCTTTAG-3′), that were previously designed on the *FtsZ* region were used [59]. The PCR mixtures were 25 µL in total, consisting of 50 ng of DNA template, 10× buffer (Thermo Fisher Scientific, Waltham, MA, USA), 25 mM of $MgCl_2$ (Thermo Fisher Scientific, Waltham, MA, USA), 2 mM of dNTPs (GeneAll, Korea), 300 nM of each primer, and five units of *Taq* DNA Polymerase (Thermo Fisher Scientific, Waltham, MA, USA). The thermal profiles used were as described previously. The positive control for *Wolbachia* PCR used was the DNA of *Wolbachia* supergroup B in *Aedes albopictus* mosquitoes, while double-distilled water was used as a negative control. The PCR products were examined using 1.5% agarose gel electrophoresis, stained with ethidium bromide, and visualized under ultraviolet transilluminator Quantity One Quantification Analysis Software Version 4.5.2 (Gel DocEQ System; Bio-Rad, Hercules, CA, USA).

4.7. Cloning and Sequencing

All positive PCR amplicons were ligated into pGEM-T Easy Vector (Promega, Madison, WI, USA), and the recombinant plasmids were used to transform a competent *Escherichia coli* DH5α strain.

Transformed cells were cultured and recombinant plasmids were then extracted using Invisorb® Spin Plasmid Mini kit (STRATEC Molecular, Berlin, Germany) following the manufacturer's instructions. Plasmids were sequenced by a commercial service at AITBIOTECH, Singapore and MACROGEN, Korea.

4.8. Sequence Analysis and Phylogenetic Tree Construction

The sequences were aligned using BioEdit Sequence Alignment Editor Version 7.2.5 [60]. The phylogenetic trees were constructed using the maximum-likelihood method with IQ-TREE on the IQ-TREE web server (http://iqtree.cibiv.univie.ac.at/) with 1000 ultrafast bootstrap replicates. The best-fit model of substitution was found using the auto function on the IQ-TREE web server [61]. The phylogenetic tree was finally viewed and edited with FigTree v1.4.4 software (http://tree.bio.ed.ac.uk/software/figtree/).

5. Conclusions

Between the aforementioned tested PCR methods, the *COI*-based method was more suitable for diagnosing canine filaria than the *ITS1*-based method. This *COI*-PCR method could differentiate *D. immitis*, *B. malayi*, and *B. pahangi* by their amplicon sizes in a single-tube PCR. This method could be useful for the epidemiological survey of filarial infection in humans, mosquitoes, and other reservoirs in Thailand. In addition, the detection of *Wolbachia* within filaria-infected dogs in our locality shows the potential of using the bacteria as a new control trend to be done simultaneously with targeting the vector and filarial parasites.

Supplementary Materials: The following are available online at http://www.mdpi.com/2076-0817/8/3/114/s1, Figure S1: PCR amplification of the partial *COI* gene for filarial nematode, Figure S2: PCR amplification of the *ITS1* region for the filarial nematode, Figure S3: Sequence alignment of the filaria nematode based on the partial *COI* gene, Figure S4: Sequence alignment of the filaria nematode based on the *ITS1* region, Figure S5: PCR amplicons of the *FtsZ* specific to *Wolbachia*, Figure S6: Sequence alignment of *Wolbachia* bacteria based on the *FtsZ* gene, Table S1: Raw data of samples.

Author Contributions: Conceptualization, K.P. and P.S.; methodology, H.S.; software, H.S. and A.P.; validation, K.P., P.S., and H.S.; formal analysis, K.P.; investigation, A.P.; resources, S.T. and S.S.; data curation, P.S. and K.P.; writing—original draft preparation, H.S. and A.P.; writing—review and editing, K.P., P.S., A.P., and N.B.; visualization, P.S.; supervision, K.P.; project administration, H.S.; funding acquisition, K.P. and P.S.

Funding: This study was supported by the Ratchadapiseksompotch Fund (grant no. RA.62/021), the Faculty of Medicine, Chulalongkorn University, National Research Council of Thailand and Chulalongkorn University (grant no. GRB_APS11593010), and the Research Assistant Scholarship, Chulalongkorn University.

Acknowledgments: We would like to thank the staff of Vector Biology and Vector Borne Disease Research Unit, Department of Parasitology, Faculty of Medicine, Chulalongkorn University, Small Animal Teaching Hospital, Chulalongkorn University, and Faculty of Veterinary Science, Mahidol University Salaya Campus, Thailand.

Conflicts of Interest: The authors declare no conflict of interest.

References

1. World Health Organization. Lymphatic Filariasis. Available online: https://www.who.int/news-room/fact-sheets/detail/lymphatic-filariasis (accessed on 28 May 2019).
2. Kaikuntod, M.; Thongkorn, K.; Tiwananthagorn, S.; Boonyapakorn, C. Filarial worms in dogs in Southeast Asia. *Vet. Integr. Sci.* **2018**, *16*, 1–17.
3. Mak, J.W.; Yen, P.K.; Lim, K.C.; Ramiah, N. Zoonotic implications of cats and dogs in filarial transmission in Peninsular Malaysia. *Trop. Geogr. Med.* **1980**, *32*, 259–264.
4. Irwin, P.J. Companion animal parasitology: A clinical perspective. *Int. J. Parasitol. Res.* **2002**, *32*, 581–593. [CrossRef]
5. Kelly, J.D. Canine heart worm disease. In *Current Veterinary Therapy VII*; W. B. Saunders: Philadelphia, PA, USA, 1979; pp. 326–335.
6. Rishniw, M.; Barr, S.C.; Simpson, K.W.; Frongillo, M.F.; Franz, M.; Dominguez Alpizar, J.L. Discrimination between six species of canine microfilariae by a single polymerase chain reaction. *Vet. Parasitol.* **2006**, *135*, 303–314. [CrossRef]

7.	Kanjanopas, K.; Choochote, W.; Jitpakdi, A.; Suvannadabba, S.; Loymak, S.; Chungpivat, S.; Nithiuthai, S. *Brugia malayi* in a naturally infected cat from Narathiwat province, southern Thailand. *Southeast. Asian J. Trop. Med. Public Health* **2001**, *32*, 585–587.

8.	Kamyingkird, K.; Junsiri, W.; Chimnoi, W.; Kengradomkij, C.; Saengow, S.; Sangchuto, K.; Ka jeerum, W.; Pangjai, D.; Nimsuphan, B.; Inpankeaw, T.; et al. Prevalence and risk factors associated with *Dirofilaria immitis* infection in dogs and cats in Songkhla and Satun provinces, Thailand. *Agric. Nat. Resour.* **2017**, *51*, 299–302. [CrossRef]

9.	Wongkamchai, S.; Nochotea, H.; Foongladda, S.; Dekumyoy, P.; Thammapalo, S.; Boitanoa, J.J.; Choochotee, W. A high resolution melting real time PCR for mapping of filaria infection in domestic cats living in brugian filariosis-endemic areas. *Vet. Parasitol.* **2014**, *201*, 120–127. [CrossRef]

10.	Simón, F.; Morchón, R.; González-Miguel, J.; Marcos-Atxutegi, C.; Siles-Lucas, M. What is new about animal and human dirofilariosis? *Trends. Parasitol.* **2009**, *25*, 404–409. [CrossRef]

11.	Ichimori, K.; King, J.D.; Engels, D.; Yajima, A.; Mikhailov, A.; Lammie, P.; Ottesen, E.A. Global programme to eliminate lymphatic filariasis: The processes underlying programme success. *PLoS. Negl. Trop. Dis.* **2014**, *8*, e3328. [CrossRef]

12.	World Health Organization. *Essential Medicines Donated to Control, Eliminate and Eradicate Neglected Tropical Diseases*; World Health Organization: Geneva, Switzerland, 2017.

13.	World Health Organization. *Lymphatic Filariasis: A Handbook of Practical Entomology for National Lymphatic Filariasis Elimination Programmes*; World Health Organization: Geneva, Switzerland, 2013.

14.	Famakinde, D.O. Mosquitoes and the Lymphatic Filarial Parasites: Research Trends and Budding Roadmaps to Future Disease Eradication. *Trop. Med. Infect. Dis.* **2018**, *43*, 4. [CrossRef]

15.	Pradatsundarasar, A. *Dirofilaria* infection in man: Report of a case. *J. Med. Assoc. Thailand* **1955**, *38*, 378–379.

16.	Jariya, P.; Sucharit, S. *Dirofilaria repens* from the eyelid of a woman in Thailand. *Am. J. Trop. Med. Hyg.* **1983**, *32*, 1456–1457. [CrossRef]

17.	Sukudom, P.; Phumee, A.; Siriyasatien, P. First report on subconjunctival dirofilariasis in Thailand caused by a *Dirofilaria* sp. closely related to *D. hongkongensis*. *Acad. J. Sci. Res.* **2018**, *6*, 114–116. [CrossRef]

18.	Taylor, M.J.; Bandi, C.; Hoerauf, A. *Wolbachia* bacterial endosymbionts of filarial nematodes. *Adv. Parasitol.* **2005**, *60*, 245–284. [CrossRef]

19.	Taylor, M.J.; Voronin, D.; Johnston, K.L.; Ford, L. *Wolbachia* filarial interactions. *Cell. Microbiol.* **2013**, *15*, 520–526. [CrossRef]

20.	Bandi, C.; Trees, A.J.; Brattig, N.W. *Wolbachia* in filarial nematodes: Evolutionary aspects and implications for the pathogenesis and treatment of filarial diseases. *Vet. Parasitol.* **2001**, *98*, 215–238. [CrossRef]

21.	Werren, J.H.; Baldo, L.; Clark, M.E. *Wolbachia*: Master manipulators of invertebrate biology. *Nat. Rev. Microbiol.* **2008**, *6*, 741–751. [CrossRef]

22.	Clark, E.L.; Karley, A.J.; Hubbard, S.F. Insect endosymbionts: Manipulators of insect herbivore trophic interactions? *Protoplasma* **2010**, *244*, 25–51. [CrossRef]

23.	Slatko, B.E.; Taylor, M.J.; Foster, J.M. The *Wolbachia* endosymbiont as an anti-filarial nematode target. *Symbiosis* **2010**, *51*, 55–65. [CrossRef]

24.	Pfarr, K.M.; Hoerauf, A.M. Antibiotics which target the *Wolbachia* endosymbionts of filarial parasites: A new strategy for control of filariasis and amelioration of pathology. *Mini. Rev. Med. Chem.* **2006**, *6*, 203–210. [CrossRef]

25.	Ravindran, R.; Varghese, S.; Nair, S.N.; Balan, V.M.; Lakshmanan, B.; Ashruf, R.M.; Kumar, S.S.; Gopalan, A.K.; Nair, A.S.; Malayil, A.; et al. Canine filarial infections in a human *Brugia malayi* endemic area of India. *Biomed. Res. Int.* **2014**, *2014*, 630160. [CrossRef]

26.	Megat Abd Rani, P.A.; Irwin, P.J.; Gatne, M.; Coleman, G.T.; McInnes, L.M.; Traub, R.J. A survey of canine filarial diseases of veterinary and public health significance in India. *Parasit. Vectors* **2010**, *3*, 30. [CrossRef]

27.	Saseendranath, M.R.; Varghese, C.G.; Jayakumar, K.M. Incidence of canine dirofilariosis in Trichur, Kerala. *Indian J. Vet. Med.* **1986**, *6*, 139.

28.	Otranto, D.; Dantas-Torres, F.; Breitschwerdt, E.B. Managing canine vector-borne diseases of zoonotic concern: Part one. *Trends Parasitol.* **2009**, *25*, 157–163. [CrossRef]

29.	Rosenblatt, J.E. Laboratory diagnosis of infections due to blood and tissue parasites. *Clin. Infect. Dis.* **2009**, *49*, 1103–1108. [CrossRef]

30. Ricciardi, A.; Ndao, M. Diagnosis of Parasitic Infections: What's Going On? *J. Biomol. Screen.* **2015**, *20*, 6–21. [CrossRef]

31. Nuchprayoon, S.; Junpee, A.; Poovorawan, Y.; Scott, A.L. Detection and differentiation of filarial parasites by universal primers and polymerase chain reaction-restriction fragment length polymorphism analysis. *Am. J. Trop. Med. Hyg.* **2005**, *73*, 895–900. [CrossRef]

32. Magnis, J.; Lorentz, S.; Guardone, L.; Grimm, F.; Magi, M.; Naucke, T.J.; Deplazes, P. Morphometric analyses of canine blood microfilariae isolated by the Knott's test enables *Dirofilaria immitis* and *D. repens* species-specific and Acanthocheilonema (syn. Dipetalonema) genus-specific diagnosis. *Parasit. Vectors* **2013**, *6*, 48. [CrossRef]

33. Hoch, H.; Strickland, K. Canine and feline dirofilariasis: Life cycle, pathophysiology, and diagnosis. *Compend. Contin. Educ. Vet.* **2008**, *30*, 133–140.

34. Casiraghi, M.; Bazzocchi, C.; Mortarino, M.; Ottina, E.; Genchi, C. A simple molecular method for discriminating common filarial nematodes of dogs (*Canis familiaris*). *Vet. Parasitol.* **2006**, *141*, 368–372. [CrossRef]

35. Wang, S.; Zhang, N.; Zhang, Z.; Wang, D.; Yao, Z.; Zhang, H.; Ma, J.; Zheng, B.; Ren, H.; Liu, S. Prevalence of *Dirofilaria immitis* infection in dogs in Henan province, central China. *Parasite* **2016**, *23*, 43. [CrossRef]

36. Byeon, K.H.; Kim, B.J.; Kim, S.M.; Yu, H.S.; Jeong, H.J.; Ock, M.S. A serological survey of *Dirofilaria immitis* infection in pet dogs of Busan, Korea, and effects of chemoprophylaxis. *Korean J. Parasitol.* **2007**, *45*, 27–32. [CrossRef]

37. Khedri, J.; Radfar, M.H.; Borji, H.; Azizzadeh, M.; Akhtardanesh, B. Canine Heartworm in Southeastern of Iran with Review of disease distribution. *Iran J. Parasitol.* **2014**, *9*, 560–567.

38. Alho, A.M.; Landum, M.; Ferreira, C.; Meireles, J.; Goncalves, L.; de Carvalho, L.M.; Belo, S. Prevalence and seasonal variations of canine dirofilariosis in Portugal. *Vet. Parasitol.* **2014**, *206*, 99–105. [CrossRef]

39. Bacsadi, A.; Papp, A.; Szeredi, L.; Toth, G.; Nemes, C.; Imre, V.; Tolnai, Z.; Szell, Z.; Sreter, T. Retrospective study on the distribution of *Dirofilaria immitis* in dogs in Hungary. *Vet. Parasitol.* **2016**, *220*, 83–86. [CrossRef]

40. Tiawsirisup, S.; Thanapaisarnkit, T.; Varatorn, E.; Apichonpongsa, T.; Bumpenkiattikun, N.; Rattanapuchpong, S.; Chungpiwat, S.; Sanprasert, V.; Nuchprayoon, S. Canine Heartworm (*Dirofilaria immitis*) Infection and Immunoglobulin G Antibodies Against *Wolbachia* (Rickettsiales: Rickettsiaceae) in Stray Dogs in Bangkok, Thailand. *Thai. J. Vet. Med.* **2015**, *40*, 165–170.

41. Ambily, V.R.; Pillai, U.N.; Arun, R.; Pramod, S.; Jayakumar, K.M. Detection of human filarial parasite *Brugia malayi* in dogs by histochemical staining and molecular techniques. *Vet. Parasitol.* **2011**, *181*, 210–214. [CrossRef]

42. Chungpivat, S.; Taweethavonsawat, P. The differentiation of microfilariae in dogs and cats using Giemsa's staining and the detection of acid phosphatase activity. *J. Thai Vet. Pract.* **2008**, *20*, 47–55.

43. Thanchomnang, T.; Intapan, P.M.; Chungpivat, S.; Lulitanond, V.; Maleewong, W. Differential detection of *Brugia malayi* and *Brugia pahangi* by real-time fluorescence resonance energy transfer PCR and its evaluation for diagnosis of *B. pahangi*-infected dogs. *Parasitol. Res.* **2010**, *106*, 621–625. [CrossRef]

44. Chansiri, K.; Tejangkura, T.; Kwaosak, P.; Sarataphan, N.; Phantana, S.; Sukhumsirichart, W. PCR based method for identification of zoonostic *Brugia malayi* microfilariae in domestic cats. *Mol. Cell. Probes* **2002**, *16*, 129–135. [CrossRef]

45. Oh, I.Y.; Kim, K.T.; Sung, H.J. Molecular Detection of *Dirofilaria immitis* Specific Gene from Infected Dog Blood Sample Using Polymerase Chain Reaction. *Iran. J. Parasitol.* **2017**, *12*, 433–440.

46. Lo, N.; Casiraghi, M.; Salati, E.; Bazzocchi, C.; Bandi, C. How many *Wolbachia* supergroups exist? *Mol. Biol Evol.* **2002**, *19*, 341–346. [CrossRef]

47. Casiraghi, M.; Bordenstein, S.R.; Baldo, L.; Lo, N.; Beninati, T.; Wernegreen, J.J.; Werren, J.H.; Bandi, C. Phylogeny of *Wolbachia pipientis* based on gltA, groEL and ftsZ gene sequences: Clustering of arthropod and nematode symbionts in the F supergroup, and evidence for further diversity in the *Wolbachia* tree. *Microbiology* **2005**, *151*, 4015–4022. [CrossRef]

48. Bordenstein, S.R.; Paraskevopoulos, C.; Hotopp, J.C.; Sapountzis, P.; Lo, N.; Bandi, C.; Tettelin, H.; Werren, J.H.; Bourtzis, K. Parasitism and mutualism in *Wolbachia*: What the phylogenomic trees can and cannot say. *Mol. Biol. Evol.* **2009**, *26*, 231–241. [CrossRef]

49. Baldo, L.; Werren, J.H. Revisiting *Wolbachia* supergroup typing based on WSP: Spurious lineages and discordance with MLST. *Curr. Microbiol.* **2007**, *55*, 81–87. [CrossRef]

50. Bandi, C.; Anderson, T.J.; Genchi, C.; Blaxter, M.L. Phylogeny of *Wolbachia* in filarial nematodes. *Proc. Biol. Sci.* **1998**, *265*, 2407–2413. [CrossRef]

51. Darby, A.C.; Armstrong, S.D.; Bah, G.S.; Kaur, G.; Hughes, M.A.; Kay, S.M.; Koldkjær, P.; Rainbow, L.; Radford, A.D.; Blaxter, M.L.; et al. Analysis of gene expression from the *Wolbachia* genome of a filarial nematode supports both metabolic and defensive roles within the symbiosis. *Genome Res.* **2012**, *22*, 2467–2477. [CrossRef]

52. Lefoulon, E.; Bain, O.; Makepeace, B.L.; d'Haese, C.; Uni, S.; Martin, C.; Gavotte, L. Breakdown of coevolution between symbiotic bacteria *Wolbachia* and their filarial hosts. *PeerJ* **2016**, *4*, e1840. [CrossRef]

53. Fenn, K.; Blaxter, M. Quantification of *Wolbachia* bacteria in *Brugia malayi* through the nematode lifecycle. *Mol. Biochem. Parasitol.* **2004**, *137*, 361–364. [CrossRef]

54. McGarry, H.F.; Egerton, G.L.; Taylor, M.J. Population dynamics of *Wolbachia* bacterial endosymbionts in *Brugia malayi*. *Mol. Biochem. Parasitol.* **2004**, *135*, 57–67. [CrossRef]

55. Nelson, G.S. The identification of infective filarial larvae in mosquitoes: With a note on the species found in "wild" mosquitoes on the Kenya coast. *J. Helmintho.* **1959**, *33*, 233–256. [CrossRef]

56. Yen, P.K.F.; Zaman, V.; Mak, J.W. Identification of some common infective filarial larvae in Malaysia. *J. Helminthol.* **1982**, *56*, 69–80. [CrossRef]

57. Orihel, T.C.; Ash, L.R.; Ramachandran, C.P.; Ottesen, E.A. *Bench Aids for the Diagnosis of Filarial Infections*; World Health Organization: Geneva, Switzerland, 1997.

58. Casiraghi, M.; Anderson, T.J.; Bandi, C.; Bazzocchi, C.; Genchi, C. A phylogenetic analysis of filarial nematodes: Comparison with the phylogeny of *Wolbachia* endosymbionts. *Parasitology* **2001**, *122*, 93–103. [CrossRef]

59. Turba, M.E.; Zambon, E.; Zannoni, A.; Russo, S.; Gentilini, F. Detection of *Wolbachia* DNA in blood for diagnosing filaria-associated syndromes in cats. *J. Clin. Microbiol.* **2012**, *50*, 2624–2630. [CrossRef]

60. Hall, T.A. BioEdit: A user-friendly biological sequence alignment editor and analysis program for Windows 95/98/NT. *Nucleic. Acids. Symp. Ser.* **1999**, *41*, 95–98.

61. Trifinopoulos, J.; Nguyen, L.T.; von Haeseler, A.; Minh, B.Q. W-IQ-TREE: A fast online phylogenetic tool for maximum likelihood analysis. *Nucleic. Acids. Res.* **2016**, *44*, W232–W235. [CrossRef]

Article

Analysis of Environmental DNA and Edaphic Factors for the Detection of the Snail Intermediate Host *Oncomelania hupensis quadrasi*

Fritz Ivy C. Calata [1], Camille Z. Caranguian [1], Jillian Ela M. Mendoza [1], Raffy Jay C. Fornillos [2,3,*], Ian Kim B. Tabios [4], Ian Kendrich C. Fontanilla [2,3,*], Lydia R. Leonardo [2,5], Louie S. Sunico [6], Satoru Kawai [7], Yuichi Chigusa [7], Mihoko Kikuchi [8], Megumi Sato [9], Toshifumi Minamoto [10], Zenaida G. Baoanan [1] and Marcello Otake Sato [7,*]

1   Department of Biology, College of Science, University of the Philippines Baguio, Governor Pack Road, Baguio City 2600, Philippines; fccalata@up.edu.ph (F.I.C.C.); caranguiancamille@gmail.com (C.Z.C.); jmmendoza8@up.edu.ph (J.E.M.M.); zgbaoanan@up.edu.ph (Z.G.B.)
2   DNA Barcoding Laboratory, College of Science, National Science Complex, University of the Philippines Diliman, Quezon City 1101, Philippines; lydialeonardo1152@gmail.com
3   Natural Sciences Research Institute, College of Science, National Science Complex, University of the Philippines Diliman, Quezon City 1101, Philippines
4   College of Medicine, University of the Philippines Manila, Pedro Gil St. Ermita, Manila 1000, Philippines; iankimbasastabios@gmail.com
5   Graduate School, University of the East Ramon Magsaysay Memorial Medical Center, 64 Aurora Blvd., Quezon City 1100, Philippines
6   Rural Health Unit, Municipal Health Office, Gonzaga, Cagayan Valley 3515, Philippines; gacuscusmark@gmail.com
7   Department of Tropical Medicine and Parasitology, Dokkyo Medical University, 880 Kitakobayashi, Mibu-machi, Shimotsuga-gun, Tochigi 321-0293, Japan; skawai@dokkyomed.ac.jp (S.K.); ychigusa@dokkyomed.ac.jp (Y.C.)
8   Department of Immunogenetics, Institute of Tropical Medicine, Nagasaki University, 1-12-4 Sakamoto, Nagasaki 852-8523, Japan; mkikuchi@nagasaki-u.ac.jp
9   Graduate School of Health Sciences, Niigata University 2-746 Asahimachi-dori, Chuo-ku, Niigata 951-8518, Japan; satomeg@clg.niigata-u.ac.jp
10   Graduate School of Human Development and Environment, Kobe University, 3-11, Tsurukabuto, Nada-ku, Kobe 657-8501, Japan; minamoto@people.kobe-u.ac.jp
*   Correspondence: rcfornillos@up.edu.ph (R.J.C.F.); icfontanilla@up.edu.ph (I.K.C.F.); marcello@dokkyomed.ac.jp (M.O.S.)

Received: 25 June 2019; Accepted: 10 September 2019; Published: 23 September 2019

Abstract: Background: The perpetuation of schistosomiasis japonica in the Philippines depends to a major extent on the persistence of its intermediate host *Oncomelania hupensis quadrasi*, an amphibious snail. While the malacological survey remains the method of choice in determining the contamination of the environment as evidenced by snails infected with schistosome larval stages, an emerging technology known as environmental DNA (eDNA) detection provides an alternative method. Previous reports showed that *O. hupensis quadrasi* eDNA could be detected in water, but no reports have been made on its detection in soil. Methods: This study, thus focused on the detection of *O. hupensis quadrasi* eDNA from soil samples collected from two selected schistosomiasis-endemic barangays in Gonzaga, Cagayan Valley using conventional and TaqMan-quantitative (qPCR) PCRs. Results: The results show that qPCR could better detect *O. hupensis quadrasi* eDNA in soil than the conventional method. In determining the possible distribution range of the snail, basic edaphic factors were measured and correlated with the presence of eDNA. The eDNA detection probability increases as the pH, phosphorous, zinc, copper, and potassium content increases, possibly indicating the conditions in the environment that favor the presence of the snails. A map was generated to show the probable extent of the distribution of the snails away from the body of the freshwater. Conclusion: The information generated from this study could be used to determine snail habitats that could be possible hotspots of transmission and should, therefore, be targeted for snail control or be fenced off from human and animal contact or from the contamination of feces by being a dumping site for domestic wastes.

Keywords: *Oncomelania hupensis quadrasi*; schistosomiasis japonica; environmental DNA; edaphic factors; snail surveillance

---

1. Introduction

Schistosomiasis, a snail-borne parasitic infection, is one of the most important neglected tropical diseases (NTDs) that continues to prevail and remains a significant public health problem in 76 tropical and subtropical countries [1–4]. The persistence of schistosomes depends highly on the continued presence of certain species of snails that serve as intermediate hosts [5–8]. There are seven major species of schistosomes that infect man but one species in particular, *Schistosoma japonicum*, is the most virulent and is considered a true zoonosis [7,9–14]. Mammals like cattle, water buffalo, goats, cats, dogs, pigs, and rodents serve as reservoir hosts [14,15]. *S. japonicum* infects an unsuspecting host via skin penetration if exposed in cercariae-contaminated freshwaters [5,6,14].

Oncomelania hupensis is a gastropod species serving as the intermediate host of *S. japonicum* in at least five countries in east and southeast Asia including the Philippines [8,16]. Currently, there are nine recognized subspecies of *O. hupensis* with taxonomic subspecies classification assigned according to their region of endemicity [16]. Four subspecies occur in China, two in Taiwan, and a single subspecies is observed each in Japan, the Philippines (*O. hupensis quadrasi*), and Indonesia [16–19]. However, recent studies based on molecular data refute most of the subspecies designation based on the geographical distribution [19].

Among the nine *O. hupensis* subspecies, *O. hupensis quadrasi* is the most amphibious [16]. It is capable of thriving both in aquatic freshwater environments as well as in muddy to dry soil areas. Under prolonged conditions of drought, *O. hupensis quadrasi* closes its operculum, burrows into the soil, and aestivates until such time when moisture returns through rain or floods [20]. This snail can thrive in both natural and man-made freshwater habitats and could be dispersed through flooding. Snail colonies in established sites are waterlogged and well-shaded [21], and their distribution in an area was reported to be influenced by breeding and survival through the variation of the soil pH and calcium [22]. Thick vegetation that provides shade and anchorage creates a suitable microclimate for the snail to thrive and reproduce [16,23,24].

Endemicity of schistosomiasis used to be observed only in provinces experiencing wet season all throughout the year until the discovery of new endemic foci in 2002 in Gonzaga, Cagayan Valley located up north (Figure 1A), an area known for summer temperatures reaching as high as 40 °C and extremely dry conditions. This discovery shows a potential adaptation of the snail to such conditions [20,25].

Schistosomiasis control in the Philippines has a long history dating as far back as 1951 when the Division of Schistosomiasis was created by the Ministry of Health [8,9,12]. Control of *O. hupensis quadrasi* snails (Figure 1B) was mainly through environmental modification, such as filling up of waterlogged areas and cementing dikes and canals or through chemical mollusciciding, which were all difficult to sustain because of the high cost and the huge demand for manpower, not to mention the very extensive areas of potential snail habitats to cover [20,26]. The shift from snail control to morbidity control was made in the early 1980s with the discovery of a cheap and effective drug, praziquantel, for all forms of schistosomiasis. Since then, mass drug administration using praziquantel has been the cornerstone of schistosomiasis control and prevention by the Philippine Department of Health (DOH). However, schistosomiasis in the Philippines is still endemic in 12 regions, 28 provinces, 14 cities,

203 municipalities, and 1,593 barangays (villages) wherein 12.4 million Filipinos are at risk of infection, and 3.4 million are directly exposed [8].

Schistosomiasis surveillance is performed to monitor progress of control programs, especially when prevalence levels have gone down to elimination levels. One specific conventional surveillance technique is the malacological survey where *O. hupensis quadrasi* snails are collected, crushed, and examined for the presence of the characteristic furcocercous cercariae of schistosomes [8,27]. The presence of infected snails is an indication of environmental contamination by fecal matter containing schistosome eggs in the freshwater [7]. Though this method is cost-effective, the demand for expertise in correctly identifying the snails in their natural habitat and for the mobilization of huge manpower during surveys, and the inaccessibility of some snail sites make the malacological survey a huge task to undertake [8,26,27].

The use of environmental DNA (eDNA) has been demonstrated to be useful in tracking *S. japonicum* in water through quantitative PCR (qPCR) [28,29]. It was also successfully applied to other schistosome species, such as *S. mansoni* in field samples for better surveillance [30,31]. In this study, *O. hupensis quadrasi* eDNA was detected from soil samples. The use of this type of DNA source material utilizing soil samples could be a useful tool for snail surveillance due to the amphibious nature of the snail.

A   B

Figure 1. (**A**) Map showing Gonzaga, Cagayan Valley in the Philippines [32]; (**B**) *O. hupensis quadrasi*, the snail intermediate host of *S. japonicum* in the Philippines, collected from Gonzaga, Cagayan Valley. The upper panel shows the aboral side and the lower panel shows the oral side where the snail's operculum could be seen.

In this study, detection of *O. hupensis quadrasi* was performed using the eDNA detection technique from soil samples collected from Gonzaga, Cagayan Valley (Figure 1A). eDNA detection was compared from snail sites harboring *O. hupensis quadrasi* and adjacent areas where no snails were observed through the classical malacological survey. Complementary, edaphic factors with putative influence in the distribution of *O. hupensis quadrasi* were investigated.

2. Results

2.1. Description of Collection Sites and Malacological Survey

The collection sites were initially determined by the presence of signages indicating infested areas (Figure 2A). A total of 27 sampling points were established: 9 from Purok 3 and 6 from Purok 5, Tapel, and 12 from Purok 4, Magrafil (Figure 2B). The sampling points positive for live *O. hupensis quadrasi* based on the malacological survey were indicated as Actual Snail Site (ASS) while two other sampling points 1 meter away from ASS were designated as Potential Snail Site (PSS). The actual snail sites were located in the map using Quantum Geographic Information System (qGIS v3.6.0) (Figure 2B). Please refer to the section on Materials and Methods for details.

The sampling areas are shown in Figure 2C. Purok 3, Tapel was a flat grassy field with swampy parts and was adjacent to a rice field. *O. hupensis quadrasi* populations were found to be randomly distributed in small rocks and on the stems of grasses in the swampy areas. The waterlogged and thick vegetation provided enough moisture and shade for the snails to establish a stable colony in the area. The snails in Purok 5, Tapel were also distributed in patches along the margins of the stream. These margins were often used as cooling and watering holes for carabaos. Farmers, along with their carabaos and pet dogs, would cross the stream on their way to their rice fields and corn plantations. Snails observed in Purok 4, Magrafil occurred in a clumped distribution and thrived along margins of a shallow stream with shady areas planted with Gabi (*Colocasia esculenta*) and Palauan (*Cyrtosperma merkusii*) while small ponds were likewise observed to support colonies of *O. hupensis quadrasi*. The collection area was contaminated with litter and fecal matter from livestock. Moreover, there were residents who bathed and washed clothes in the stream.

Figure 2. *Cont.*

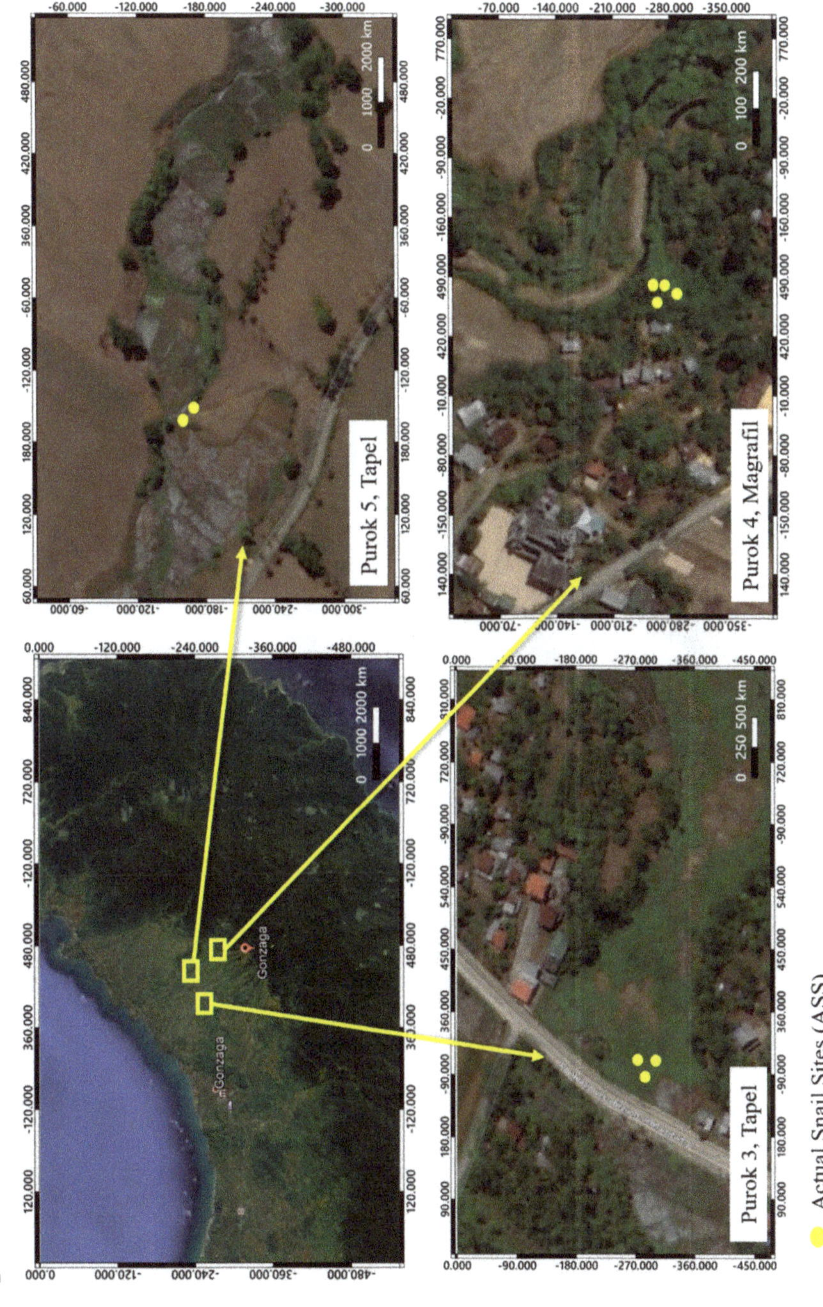

Figure 2. *Cont.*

C

Figure 2. (**A**) Signages of schistosomiasis-endemic sites in Gonzaga, Cagayan Valley; (**B**) Actual snail sites mapped using qGIS 3.6.0. visited in barangays Tapel and Magrafil for soil collection; (**C**) Sampling collection sites in Purok 3, Tapel, Purok 5, Tapel, and Purok 4, Magrafil.

2.2. eDNA Detection via Conventional PCR and qPCR

All the 81 eDNA samples were run in triplicates using both conventional PCR and qPCR. (Supplementary Material Table S1). Conventional PCR detected *O. hupensis quadrasi* eDNA in only one of the 27 sampling points (data not shown). On the other hand, qPCR detected *O. hupensis quadrasi* eDNA in 22 out of 27 (81.48%) sampling points as indicated in the Ct values and amplification plots (Supplementary Material Table S2).

Specifics of eDNA detection rate is summarized in Figure 3. In Purok 3, Tapel, all ASS (100%) had detectable eDNA, whereas all 5 out of 6 PSS were positive for eDNA (83.33%). In Purok 5, Tapel, no eDNA was detected in 2 ASS visited, whereas 2 out of 4 PSS (50%) had detectable eDNA. Moreover, eDNA was detected in all sampling areas in Purok 4, Magrafil.

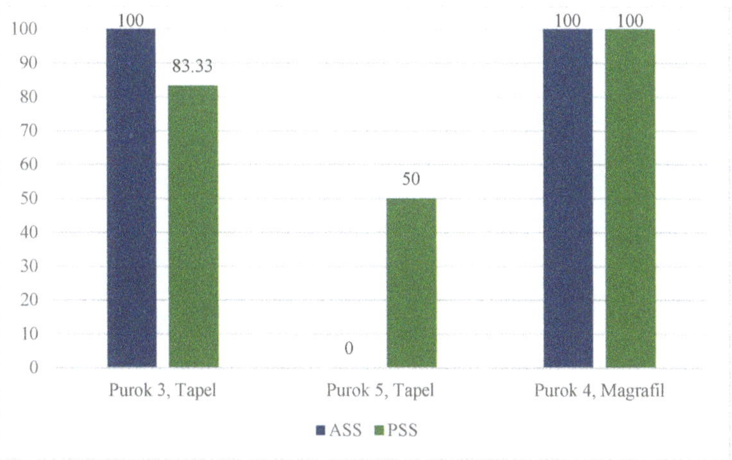

Figure 3. eDNA detection rate of *O. hupensis quadrasi* from actual snail sites (ASS) and potential snail sites (PSS) using qPCR in three Puroks from barangays Tapel and Magrafil, Gonzaga, Cagayan Valley.

3. Distribution Mapping of eDNA Positive Sites

Specific location points of ASS and PSS classified according to the results of qPCR detection of *O. hupensis quadrasi* eDNA were determined using Global Positioning System (GPS) and plotted in the map using qGIS (Figure 4A–C). Water bodies such as streams as observed during the field visit were also incorporated in the map as indicated in the legend of each map generated.

A.

Figure 4. *Cont.*

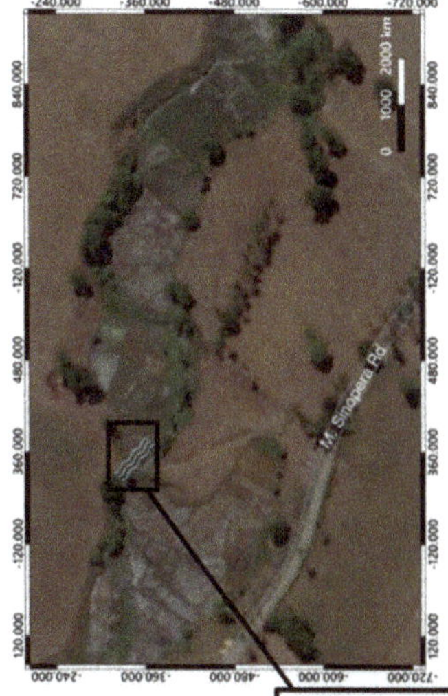

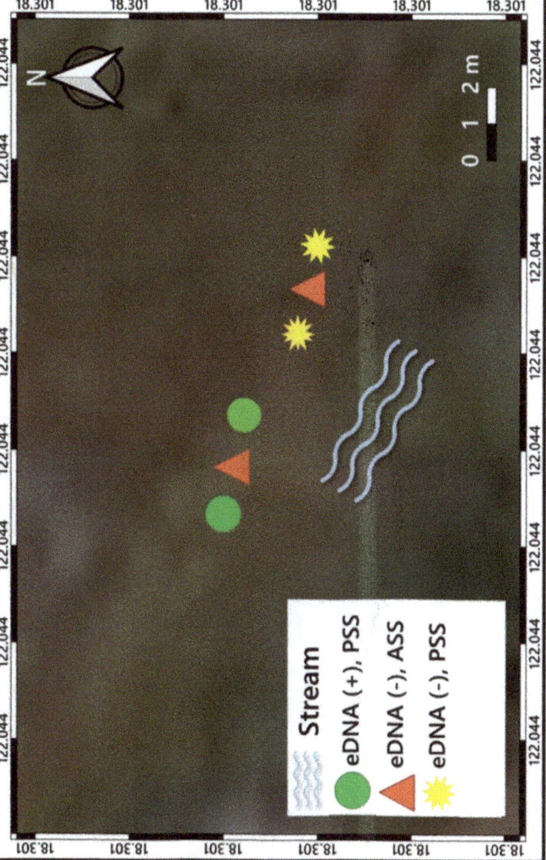

Figure 4. Cont.

C.

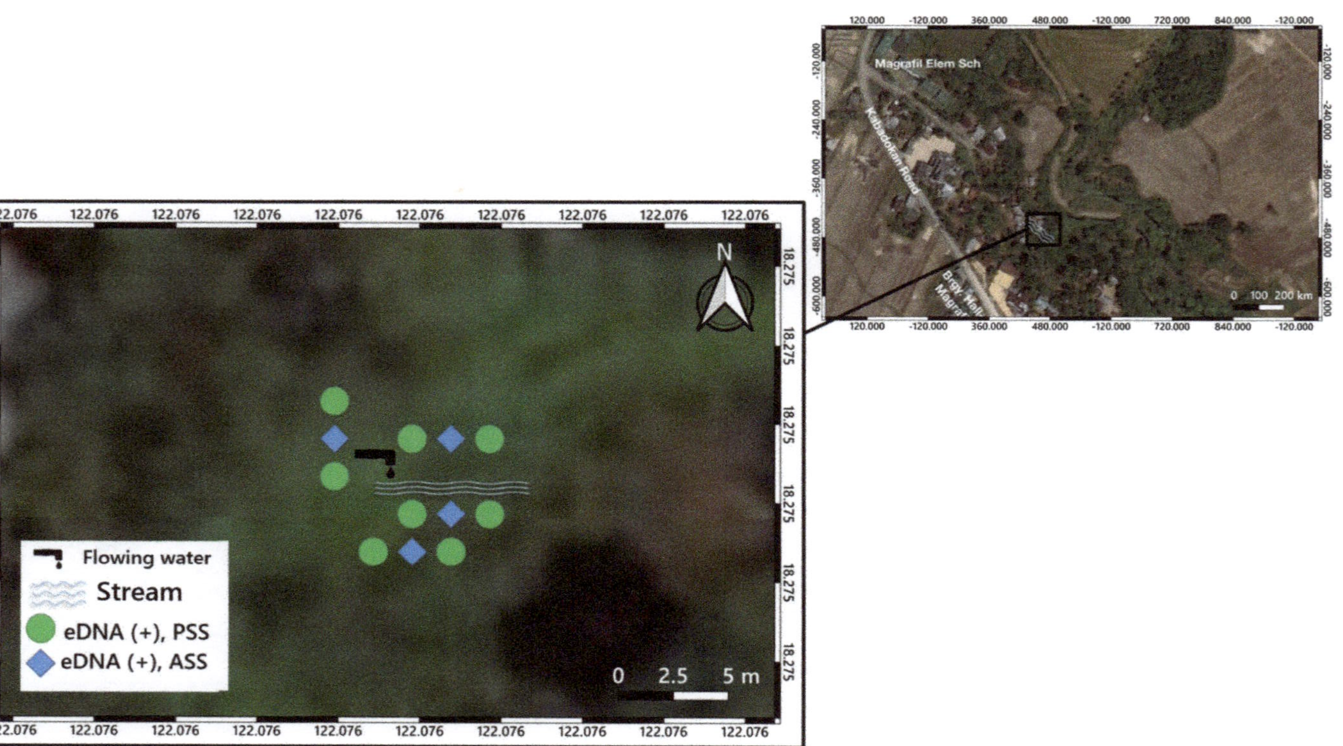

Figure 4. Actual sampling areas of ASS and PSS plotted in qGIS v.3.6.0. using recorded Global Positioning System (GPS) points either positive or negative to *O. hupensis quadrasi* eDNA of sampled areas of (**A**) Purok 3, Tapel, (**B**) Purok 5, Tapel, and (**C**) Purok 4, Magrafil.

4. Soil Edaphic Factors

Edaphic factors that were measured are summarized in Supplementary Material Table S3. These factors were correlated to the eDNA detection probability (number of positive eDNA readings in the qPCR of a sample) using Statistical Package for the Social Sciences (SPSS). There were positive correlations observed between edaphic factors, such as pH, P, K, Cu, and Zn (Figure 5A–E) in which higher detection rates were observed in samples with increasing levels of the aforementioned factors in the soil. On the other hand, no correlations were observed for temperature, Ca, Mn, N, Fe, and organic matter (data not shown).

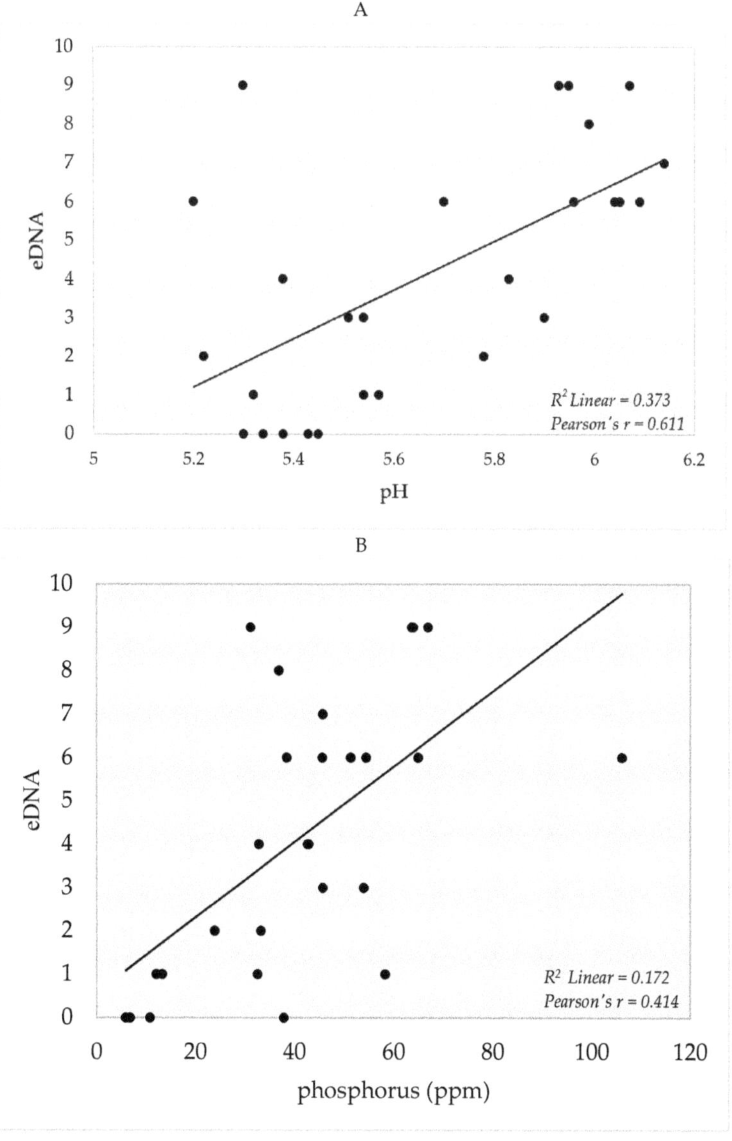

Figure 5. *Cont.*

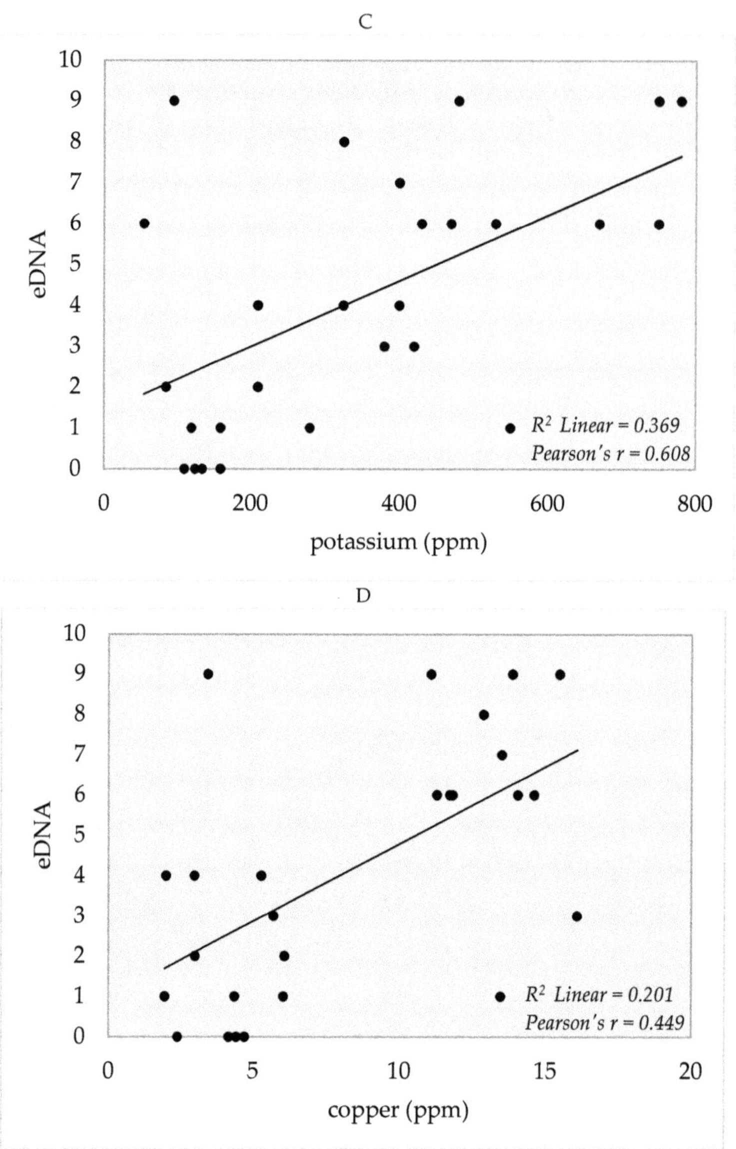

Figure 5. *Cont.*

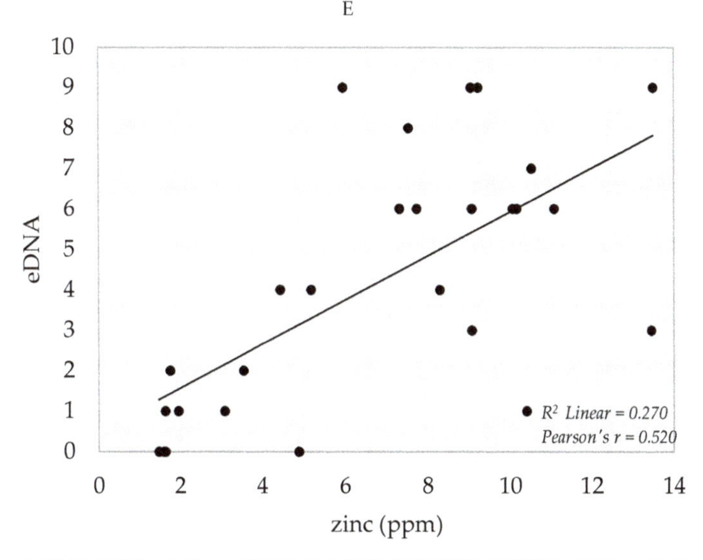

Figure 5. Positive correlation scatter plots of eDNA detection probability vs. (**A**) pH, (**B**) phosphorus, (**C**) potassium, (**D**) copper, (**E**) zinc calculated using Pearson Correlation in SPSS v. 16.

5. Discussion

5.1. Detection of eDNA of O. hupensis quadrasi from Soil Samples via Conventional PCR and qPCR

Results of eDNA detection using conventional PCR and qPCR are summarized in Supplementary Material Table S1 and Figure S1. eDNA was detected using both conventional PCR and qPCR, but detection rates in the latter were higher. The use of TaqMan-qPCR also addresses the problem of the production of non-specific fluorescence signal from non-specific amplification [27]. This is due to the high-sequence specificity of the primers and probes designed, and a signal is only generated if the binding to the target sequence of the primers and probes to the target DNA is observed in the reaction. The utilization of probes with the primers increases not only the sensitivity of the qPCR but most importantly the specificity of the reaction.

The A260/230 ratios of spectrophotometry of the soil samples confirm the presence of protein contaminants and organic compounds. In the samples, the values were extremely lower than the ideal range, which indicates that the eDNA extracts were not pure and contain significant contaminants such as proteins, carbohydrates and inorganic compounds such as salts that can inhibit PCR. The qPCR negative results observed in ASS in Tapel (Figures 3 and 4A,B) may be due to compounds acting as inhibitors during PCR which results in zero amplification even if snails are observed in these sites [33,34]. Another plausible reason could be the rapid degradation of DNA before it even settled to the sediment. It is possible to detect snails in the field, but little to no eDNA could be detected in soil sediments due to the rapid degradation processes to which unbound DNA are exposed to [35,36]. The amount of eDNA that has settled on the soil sediment is dependent on the rate of eDNA shedding of the source organism. Failure to collect the soil with the target DNA even though live *O. hupensis quadrasi* is present at the site is therefore possible. To address this problem, extensive sampling points were designed to systematically collect soil from these areas so that the probability of detecting eDNA would be higher. Nevertheless, the detectable eDNA of *O. hupensis quadrasi* confirmed the efficacy of eDNA detection through qPCR even with high carbohydrate carryover and low initial DNA concentration. As the probability of detecting eDNA and the ability to recover it varies with eDNA concentration, the inconsistencies of the detection among qPCR replicates (Supplementary Material

Table S1) are anticipated since the concentration of the initial target DNA is extremely low [36,37]. At least one positive result from the three repeated runs for a particular sample in the qPCR test was considered positive to *O. hupensis quadrasi* eDNA.

The utilization of eDNA detection will enhance the efficiency of the snail surveillance, most especially in areas where the snail density is low, and the distribution is too patchy. Using soil as material, it is also possible to detect potential snail habitats that may harbor snails even during the dry season. Since the malacological survey as a surveillance technique is done routinely in some endemic areas in the Philippines, eDNA detection could be applied as a complementary technique to confirm the presence of the snails or if the area has been inhabited by the snail since eDNA could persist for a long duration of time. Although the current study did not include the detection of *S. japonicum* using water, it is more relevant to track the parasite using water due to its high dependability on the medium to complete its life cycle [7,8,27]. Nevertheless, detection of *S. japonicum* using soil samples could also be explored since it may indicate the presence of infected snails. It can also determine if the sampled area is a site of active transmission where continuous environmental contamination is experienced through the presence of miracidia emerging from eggs introduced through the fecal matter from infected hosts to freshwater bodies. The detection of *S. japonicum* in soil samples may also provide useful geographic implications on the range of cercarial displacement in dry areas where flooding is prominent. This could also be achieved by using *S. japonicum* primers and *S. japonicum* specific probes to increase the sensitivity and specificity of the detection using qPCR. However, since *S. japonicum* is not visible to the naked eye, it would be difficult to assign sampling points for detection.

Although eDNA detection can provide an alternative method to survey snails, this method cannot assess the typical density of *O. hupensis quadrasi* and its real-time location, whether it comes from a recently sloughed-off DNA or a dead snail [38–41]. The success of using the eDNA method further depends on certain characteristics of the species of interest. For example, it is easier to detect species such as fish and amphibians with slimy skin since they release great quantities of DNA in the environment as compared to arthropods which only release small amounts [39]. In terms of *O. hupensis quadrasi*, its high dependability and contact with soil and water increase the chance of eDNA shedding, making either water or soil a good candidate material to use to detect them.

5.2. O. hupensis quadrasi eDNA Potential Distribution

In Purok 3, Tapel, all ASS and 5 PSS were positive for eDNA (Figure 4A), which indicates that even if there was absence of snails in the PSS during sampling, it could mean that *O. hupensis quadrasi* could have been in the site in the past or its eDNA could have been transported to nearby PSS. The presence of eDNA in some PSS may imply that the snails were present in the site but were missed during the malacological survey or that the eDNA could have originated from a nearby ASS as eDNA exhibits high mobility.

No eDNA was detected in all ASS and two PSS were positive with eDNA in Purok 5, Tapel (Figure 4B). The presence of *O. hupensis quadrasi* eDNA in PSS indicates a potential snail habitat. Moreover, the presence of eDNA indicates the possible distance of eDNA dispersion from the nearest source such as an actual snail site, which was used as a reference during sampling. *O. hupensis quadrasi* spawn and sometimes shed *S. japonicum* cercariae along the stream, then return to the soil substrate where they settle and possibly aestivate. In Purok 4, Magrafil, all ASS and PSS were positive for eDNA (Figure 3; Figure 4C), which may be attributed to the high abundance of snails. It can be inferred from the maps that the snail can travel 0.5 m to 6.5 m from the stream to the sampling points. In addition, it is possible that *O. hupensis quadrasi* is dispersed by livestock animals, humans, or by continuously flowing water.

It is difficult to accurately track the source and movement of DNA molecules since eDNA data relies on inference [42]. For instance, dispersion and dilution of eDNA may be affected by stream currents and wave actions [39,43]. In the case of *O. hupensis quadrasi*, the presence of eDNA may confirm the extent of snail distribution after flooding. eDNA detection is easier in species living in small isolated areas since sampling could be limited to their particular habitats. Meanwhile, eDNA detection could be difficult in species with greater means of dispersal living in big rivers or terrestrial habitats since sampling would have to be extensive. There is a high probability of positive eDNA detection if the target species is in the area recently since DNA degrades over time [39]. Thus, it is essential to have an appropriate sampling design based on whether the collected sample is near or far from the target species.

5.3. Edaphic Factors and eDNA Detection

Results of measured edaphic factors in sampling areas are summarized in Supplementary Material Table S3. There were positive correlations in the eDNA detection probability with the edaphic factors pH, P, K, Zn, and Cu (Figure 5). On the other hand, no correlations were found with temperature, N, Fe, Ca, Mn, organic matter, and organic carbon (data not shown). Knowledge of these parameters, which may regulate the local distribution of *O. hupensis quadrasi* or the persistence of the snail's DNA, is necessary to determine the parameters' effects on the distribution of the snail.

There was a moderate positive correlation ($\varrho = 0.611$) in the pH with eDNA detection (Figure 5A). Although pH values increased as the number of eDNA positive readings in qPCR increased, increasing pH may not necessarily mean that *O. hupensis quadrasi* could tolerate a very basic environment since the snail generally thrives at a range of 4.6 to 9 [22]. The positive correlation in our results, however, may imply that *O. hupensis quadrasi* could tolerate slightly acidic environments (5.20 to 6.14) or the snail's DNA is stable in these pH conditions. If the pH becomes very basic, the *O. hupensis quadrasi* productivity becomes low [22], whereas a very acidic environment promotes DNA degradation and shell erosion, which may lead to death [44].

eDNA detection had a moderate positive correlation ($\varrho = 0.414$) with phosphorus (P) (Figure 5B). The P range of 5.9 to 106 ppm may indicate that *O. hupensis quadrasi* can thrive within this range and detection of the snail's DNA is possible. Pesigan and colleagues [21] found that P concentration has no effect on the distribution of the snails in Palo, Leyte. However, the results of this study point otherwise. This is supported by a study by Johnson and colleagues [45], which found that P has an effect on the snail distribution since P can serve as a nutrient. Increased amounts of P may be due to the sewage released by nearby households in Gonzaga, which allowed the perpetuation of *O. hupensis quadrasi* and the eventual increase of detected eDNA [46].

The probability of detecting eDNA was also noted to increase with increasing potassium (K) levels, showing a moderate positive correlation ($\varrho = 0.608$) (Figure 5C). The increasing K levels may be due to the high moisture available in the soil during sampling, which was performed during the wet season. Moisture permits K leaching from crop residues (*Oriza sativa* and *C. esculenta*) to soil [47]. Thus, moisture might still be available to *O. hupensis quadrasi* despite being 0.5 to 6.5 m away from the stream (Figure 5B,C). The positive correlation between K and detected eDNA might, therefore, infer a possible tolerance of *O. hupensis quadrasi* to ecotoxic potassium-based compounds, such as potassium nitrate (KNO_3) and potassium chloride (KCl) [47].

Copper (Cu) also showed a moderate positive correlation ($\varrho = 0.449$) with number of detected eDNA (Figure 5D). The correlation may reveal a possible tolerance of the snail to increasing Cu (up to 16.08 ppm) since this element is toxic to snails with increasing amounts [48]. In other gastropod species such as *Melanoides tuberculata* and *Theodoxus niloticus*, the mean lethal concentrations to Cu are 0.14 ppm and 8.6 ppm, respectively [48,49]. A study by Moreno and McCord [50] showed that high Cu has adverse effects on DNA processing by directly interacting with the DNA, thereby altering its chemical structure [51]. Moreover, Cu-based compounds such as copper sulfate ($CuSO_4$), copper pentachlorophenate, and copper-controlled release glass (CRG) are used as molluscicides to intermediate hosts of *Schistosoma* spp. [52,53]. $CuSO_4$ specifically is known to be toxic to snails in minute doses for as low as 2 ppm, which can cause death after 48 hours of exposure [52].

Zinc (Zn) also showed a moderate positive correlation ($\varrho = 0.520$) (Figure 5E). Since compounds of Zn such as ZnO nanoparticles serve as molluscicides to *Oncomelania* snails and

Biomphalaria alexandrina [54,55], the positive correlation of zinc with eDNA detection may indicate a possible tolerance of *O. hupensis quadrasi* to rising zinc levels (up to 13.48 ppm; Supplementary Material Table S3). The mean lethal concentrations of 3.9 ppm and 12.2 ppm occur in *M. tuberculata* and *T. niloticus*. There is a high probability that zinc-based consumer products (e.g., batteries) leach out from Zn-contaminated sites to the endemic sites [56].

Temperature, Ca, Mn, N, Fe, organic matter, and organic carbon showed no significant correlation with the eDNA detected in soil. Although *O. hupensis quadrasi* is known to be sensitive to changes in temperature, the increasing temperatures did not affect eDNA detection since sampling was only performed during the wet season and only minute differences were observed in soil temperatures [21,22,57]. Calcium is critical to the survival of *O. hupensis quadrasi* as it is needed for the development of their calcareous shell [58]. Hence, increasing calcium content should result in increasing the probability of detecting the snail's eDNA. However, results in this study show the absence of a correlation between eDNA detection and calcium. It is hypothesized that the three-month gap between soil collection and soil analysis altered the initial calcium content of the samples. Similarly, a study by Pesigan et al. [21] suggests that calcium have no effect on snail distribution in Palo, Leyte. Nevertheless, based on this study, it is possible that the snail could survive or its eDNA could persist in varying concentrations of calcium (0.25–10 cmol/kg), organic carbon (0.9–3%), and organic matter (1.5–5.3%).

Furthermore, the absence of *O. hupensis quadrasi* in places appearing to be suitable for them led McMullen to conclude that changes in the environment may be responsible for the altered tolerance range of *O. hupensis quadrasi* [21,23]. Results from this study, however, do not indicate the overall general trend for the soil factors affecting *O. hupensis quadrasi*, as other environmental factors may affect the presence of *O. hupensis quadrasi* eDNA. Furthermore, the trend in the correlation of the eDNA detection probability and edaphic factors measured may change seasonally since sampling was only done for one season for this study. More extensive temporal trends in the variables measured may provide higher resolution of eDNA detection probability and edaphic factors that may affect the distribution of *O. hupensis quadrasi* in an endemic area.

6. Conclusions and Recommendations

This is a pioneering study showing the success of detecting eDNA of *O. hupensis quadrasi* in soil despite the low initial DNA concentrations of the target organism and the presence of inhibitors. eDNA was detected in 22 out of 27 sites (81.48%) of barangay Tapel and Magrafil via qPCR while only one out of 27 sampling sites (3.70%) tested positive using conventional PCR, validating the sensitivity of qPCR over conventional PCR. Distribution mapping confirmed that *O. hupensis quadrasi* could extend their habitat range or their eDNA could be dispersed from 0.5 m to 6.5 m away from the stream. This could also indicate the ability of the snails to thrive in relatively drier areas which may be far from water sources. Pearson's positive correlation showed that the detection of eDNA (number of positive readings in qPCR) increases as the pH, phosphorous, potassium, copper and zinc content increase. The correlation may infer a possible tolerance range of *O. hupensis quadrasi* or persistence of its eDNA in these conditions. Thus, the initial snail distribution provides basic information about the snail intermediate host, which could be further analyzed to mitigate the threats of schistosomiasis in endemic areas.

The limitations and benefits of eDNA detection should be evaluated when selecting a method in targeting vectors, intermediate hosts, and parasites. Seasonal changes of eDNA concentrations and biomass or the abundance through eDNA quantification in the soil could be further studied. Detection of eDNA of both the intermediate snail host and the parasite is also suggested to determine the extent of transmission of schistosomiasis in an endemic area. The effect of temperature on eDNA distribution could also be evaluated throughout the year, involving the sampling during wet and dry seasons to determine the possible tolerance range and adaptation of *O. hupensis quadrasi*. *O. hupensis nosophora* found in Japan, which used to be endemic for schistosomiasis until it was eliminated in the 1990s, experienced a drastic reduction in the wild populations due to intensive snail control measures as part of the intensive campaign to eliminate and permanently interrupt transmission of the disease [17]. The same is observed in some subspecies of *O. hupensis* in China, where intensive mollusciciding and physical environmental modification are still practiced [18].

7. Materials and Methods

7.1. Soil Sampling, Malacological Survey, and Mapping of Snail Sites

Soil sampling was conducted in January 2019 in schistosomiasis-endemic areas in Purok (zone) 3 and Purok 5 of Barangay (village) Tapel and Purok 4 of Barangay Magrafil in Gonzaga, Cagayan from actual snail sites (ASS) and potential snail sites (PSS). ASS are places with snails confirmed through the intensive malacological survey in which at least one hour was spent searching for snails. The PSS sites were established as two quadrats at one-meter distance to the left and right of ASS with the confirmed absence of snails by the malacological survey (Figure 6A).

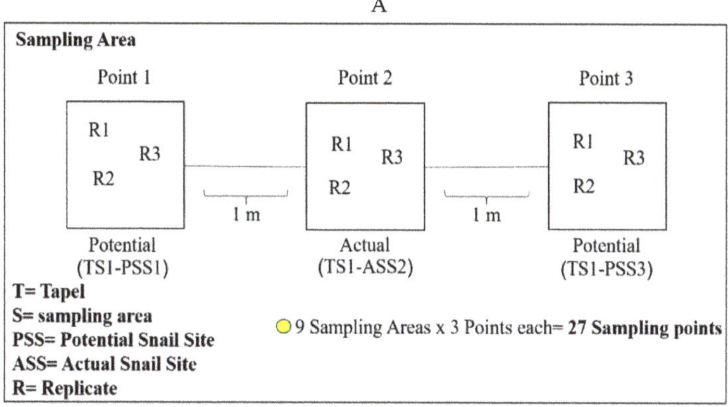

Figure 6. (**A**) Established ASS and PSS in one sampling area; (**B**) Overall sampling points and soil samples in barangay Tapel and Magrafil.

Overall, there were nine sampling areas (Figures 2B and 6A): three in Purok 3, Tapel; two in Purok 5, Tapel; and four in Purok 4, Magrafil, with nine ASS sampling points and 18 PSS sampling points for a total of 27 (Figure 6A,B). In each sampling point, three replicates of soil samples were collected accounting to a total of 81 soil samples (Figure 6B). The GPS coordinates of all sampling points of ASS and PSS, whether positive or negative to *O. hupensis quadrasi* eDNA, were also recorded and mapped (Figure 4).

To perform the malacological survey, five people were involved. Sites with actual snails were screened by intensively searching for live *O. hupensis quadrasi*. Appropriate protective clothing such as knee-level boots and gloves were used. Snails were collected with the use of forceps and were stored in properly labeled plastic cups. Photographs and GPS coordinates using a portable GPS device (Etrex H, Garmin) were taken and recorded for each of the established snail sites and were mapped using qGIS 3.6.0 (Figure 2B).

Soil samples were taken at a depth of approximately 10 cm. For each ASS and PSS sites, triplicate soil samples of about 60 g were collected and placed separately in properly labeled polyethylene bags (size 8″ × 10″) for eDNA extraction. Another 500 g sample of soil was set aside from each sampling area (Tuguegarao, Cagayan Valley, and Capas, Tarlac) for soil quality analysis. During the soil collection, the trowel used for each site was rinsed with a 10% bleach solution after every sampling to prevent cross-contamination. Shoulder-length gloves were used in soil collection to avoid exposure to the parasite.

7.2. Storage of Soil Samples and Measurement of Edaphic Factors

The soil samples were transported to the Oven Room, KA Building of the University of the Philippines Baguio for further processing. Samples were oven-dried at ~40 oC for about two weeks. Dried soil samples were submitted to the Department of Agriculture Region II and III Soils Laboratory for analyses of soil pH, phosphorus (P), potassium (K), zinc (Zn), copper (Cu), manganese (Mn), and iron (Fe), organic matter (OM), organic carbon (OC), and calcium (Ca). Soil temperature was measured in situ at the time of collection using a portable thermometer.

7.3. eDNA Extraction from Soil and Nanodrop Spectrophotometry

Eighty-one samples weighing 60 grams each were collected and allotted for eDNA extraction. Each sample was placed in a separate 500 mL sterile beaker, whereupon distilled water was added to create a suspension solution. Fifteen mL from the suspension was immediately subsampled into a 50 mL sterile conical centrifuge tube containing 33 mL of absolute ethanol and 15 mL of 3 M sodium acetate for preservation. The resulting suspension solutions were then stored in a cooler (approximately 0 °C) and were transported to the DNA Barcoding Laboratory at the Institute of Biology Laboratory in UP Diliman for eDNA extraction. The samples were thawed, homogenized, and centrifuged at 8500 rpm for 30 min. The supernatant was discarded and the sediments were homogenized and collected from the centrifuge tube. Sediment of 0.25 g was added to the Lysing Matrix E tube and subsequently subjected to the eDNA extraction protocol of the FastDNA® SPIN Kit for Soil (MP Biomedicals Europe). Extracted eDNA samples were then subjected to spectrophotometry using Nanodrop 2000 (Thermo Scientific) to check for any contamination and DNA purity. The extracted DNA samples were then stored at −20 °C until use.

7.4. Detection via Conventional PCR and qPCR

Each of the 81 extracted eDNA was tested in triplicates using conventional PCR and qPCR to determine the presence of *O. hupensis quadrasi* eDNA. The specific primers and probe to *O. hupensis quadrasi* were designed manually by the alignment of cytochrome *c* oxidase subunit 1 gene sequences of *O. h. quadrasi*, other *Oncomelania* subspecies, and related taxa from other gastropod species from the families Planorbidae, Ampullariidae, Neritidae, Achatinidae, and Thiaridae. The designed primers and probe sequences were then searched against the nucleotide sequence database using BLAST to examine other potential targets. A 187 base pair fragment of *cox1* was targeted using the forward primer OhqCOX1_22-41aF (GCATGTGAGCGGGGCTAGTA) and reverse primer OhqCOX1_189-209aR (AAGCGGAACCAATCAGTTGCC). Positive controls containing pure *cox1* gene of *O. hupensis quadrasi* were used to validate the qPCR setting and primer pair design whereas a negative control without the template was utilized to check for any contamination. For conventional PCR, a 12.50 μL reaction volume per sample was prepared by mixing 7.2 μL of RNase/DNase-free polymerase chain reaction (PCR) grade water (Ambion, Thermo Scientific), 2.5 μL 10X PCR Buffer, 0.64 μL 2.5 mM dNTP mix, 0.25 μL 25 mM MgCl$_2$, 0.38 μL μmol each of forward and reverse primers, 0.06 μL 5 U/μL Taq Polymerase (Takara-Clontech), and 1.0 μL of DNA extract with at least 35 ng/μL concentration. PCR was performed in T100™ Thermal Cycler. PCR conditions were the following: 95 °C for initial denaturation for 30 s, 95 °C for denaturation for 5 s, and 60 °C for annealing for 30 s for 50 cycles. PCR products were visualized in a 2% agarose gel (Vivantis Technologies Sdn Bhd) dissolved in 0.5X TBE and stained with 1% ethidium bromide (EtBr) run for 30 mins in 100 v in a horizontal AGE apparatus. *O. hupensis quadrasi* eDNA was also targeted through TaqMan-quantitative real-time PCR using TaqMan System technology in an Applied Biosystems 2720 Thermal Cycler Dice® Real-Time System II (TP 900). The forward and reverse primers used in the conventional PCR were also used. A 10 μL master mix was prepared by mixing 5.5 μL TaqMan Environmental Master Mix (EMM), 1.01 μL of each forward and reverse primers, 0.28 μL TaqMan custom probe (OhqCOX1_67-86P 5′-FAM-GTGCAGAGTTAGGTCAGTCCT-MGB-NFQ-3′), and 2.95 μL of extracted *O. h. quadrasi* eDNA. Samples were then transferred onto well plates and subjected under the following thermal cycling parameters: 95 °C for AmpliTaq Gold®, UP enzyme (DNA polymerase) activation under 10 min, 95 °C for denaturation for 15 s, and 60 °C for annealing under 1 min, repeated for 40 cycles (Supplementary Material Figure S1).

7.5. Statistical Analysis

Soil factors and eDNA recovery in all actual and potential snail sites and sampling points were correlated using Pearson's Correlation available in the SPSS (Statistical Package for the Social Sciences) software v. 16. Pearson's correlation was utilized to determine the possible linear association between the two variables [soil factor and eDNA detection probability (number of eDNA positive readings in qPCR for a sample)].

Supplementary Materials: The following are available online at http://www.mdpi.com/2076-0817/8/4/160/s1, Table S1. Malacological survey and conventional Polymerase Chain Reaction (PCR) and TaqMan-quantitative Polymerase Chain Reaction (qPCR) readings of the number of detected environmental DNA (eDNA). (Legend: T-Tapel, S1-Sampling Point Number, PSS1- Potential Snail Site Number, R1-Replicate Number, ASS2- Actual Snail Site Number, M- Magrafil). Table S2. qPCR of *O. hupensis quadrasi* eDNA in barangay Tapel and barangay Magrafil with corresponding Ct (cycle threshold values) and Ct threshold. Table S3. Edaphic factors of all soil samples in Barangay Tapel and Magrafil. Figure S1. TaqMan qPCR amplification plots of *O. hupensis quadrasi*, showing cycle threshold (Ct) levels as indicated by the blue horizontal line; the positive control is indicated by the pink line. The positive result is indicated by amplifications exceeding the threshold line. Each color represents a single sample.

Author Contributions: Conceptualization, R.J.C.F.; I.K.B.T.; I.K.C.F.; L.R.L.; Z.G.B.; and M.O.S.; Methodology, F.I.C.C.; C.Z.C.; J.E.M.M.; R.J.C.F.; Software, I.K.C.F.; and M.O.S.; Validation, I.K.C.F.; L.R.L.; and M.O.S.; Formal Analysis, F.I.C.C.; C.Z.C.; J.E.M.M.; and R.J.C.F.; Investigation, I.K.C.F.; L.R.L.; Z.G.B.; and M.O.S.; Resources, L.S.S.; S.K.; Y.C.; M.K.; M.S.; T.M.; Z.G.B.; M.O.S; Data Curation, F.I.C.C.; C.Z.C.; J.E.M.M.; and R.J.C.F.; Writing-Original Draft Preparation, F.I.C.C.; C.Z.C.; J.E.M.M.; and R.J.C.F.; Writing-Review & Editing, I.K.B.T.; I.K.C.F.; L.R.L.; L.S.S.; S.W.; Y.C.; M.K.; M.S.; T.M.; Z.G.B.; and M.O.S.; Visualization, F.I.C.C.; C.Z.C.; and J.E.M.M.; Supervision, R.J.C.F.; I.K.C.F.; L.R.L.; and Z.G.B.; Project Administration, R.J.C.F.; I.K.C.F.; and L.R.L.; Funding Acquisition, R.J.C.F.; I.K.B.T.; I.K.C.F.; and L.R.L.

Funding: This study was funded by the Department of Science and Technology Philippine Council for Industry, Energy and Emerging Technology Research and Development (DOST-PCIEERD) granted to Ian Kendrich C. Fontanilla (Grant No. PCIEERD-BIO-17-02) of University of the Philippines Diliman-Institute of Biology (UPD-IB) and Natural Sciences Research Institute (NSRI) in partnership with the Department of Biology, College of Science of University of the Philippines Baguio.

Acknowledgments: The researchers would like to thank DOST-PCIEERD, Diliman, and Baguio campuses of the University of the Philippines and NSRI for supporting the study. We would also like to express our appreciation to the Region II offices of the Department of Health, Region II and III offices of the Department of Agriculture, Bureau of Soils and Water Management for the provision of soil analysis. The same gratitude also goes to the Local Government Unit, Municipal Health Office, and the Rural Health Unit of Gonzaga, Cagayan Valley, and all personnel who guided us during the sample collection and field survey. The researchers would also like to thank all collaborating universities and agencies in Japan for the resources and extension of expertise.

Conflicts of Interest: The authors declare no conflict of interests.

References

1. King, C.H.; Dickman, K.; Tisch, D.J. Reassessment of the cost of chronic helmintic infection: A meta-Analysis of disability-Related outcomes in endemic schistosomiasis. *Lancet* **2005**, *365*, 1561–1569. [CrossRef]

2. Steinmann, P.; Keiser, J.; Bos, R.; Tanner, M.; Utzinger, J. Schistosomiasis and water resources development: Systematic review, meta-Analysis, and estimates of people at risk. *Lancet Infect. Dis.* **2006**, *6*, 411–425. [CrossRef]

3. Finkelstein, J.L.; Schleinitz, M.D.; Carabin, H.; McGarvey, S.T. Decision-Model estimation of the age-Specific disability weight for schistosomiasis japonica: A systematic review of the literature. *PLoS Negl. Trop. Dis.* **2008**, *2*, 158. [CrossRef] [PubMed]

4. Zhou, Y.; Zheng, H.; Chen, X.; Zhang, L.; Wang, K.; Guo, J.; Huang, Z.; Zhang, B.; Huang, W.; Jin, K.; et al. The Schistosoma japonicum genome reveals features of host-Parasite interplay. *Nature* **2009**, *460*, 345.

5. Gryseels, B.; Polman, K.; Clerinx, J.; Kestens, L. Human schistosomiasis. *Lancet* **2006**, *368*, 1106–1118. [CrossRef]

6. Colley, D.G.; Bustinduy, A.L.; Secor, W.E.; King, C.H. Human schistosomiasis. *Lancet* **2014**, *383*, 2253–2264. [CrossRef]

7. Olveda, D.U.; Li, Y.; Olveda, R.M.; Lam, A.K.; McManus, D.P.; Chau, T.N.P.; Harn, D.A.; Williams, G.M.; Gray, D.J.; Ross, A.G.P. Bilharzia in the Philippines: Past, present, and future. *Int. J. Infect. Dis.* **2014**, *18*, 52–56. [CrossRef] [PubMed]

8. Leonardo, L.; Chigusa, Y.; Kikuchi, M.; Kato-Hayashi, N.; Kawazu, S.I.; Angeles, J.M.; Fontanilla, I.K.; Tabios, I.K.; Moendeg, K.; Goto, Y.; et al. Schistosomiasis in the Philippines: Challenges and some successes in control. *Southeast Asian J. Trop. Med. Public Health* **2016**, *47*, 651–666.

9. Blas, B. Handbook for the control of Schistosomiasis japonica. *Monogr. Schist Jpn. Infect. Philipp.* **1991**, *72*, 745–753.

10. Ross, A.G.P.; Sleigh, A.C.; Li, Y.; Davis, G.M.; Williams, G.M.; Jiang, Z.; Feng, Z.; Manus, D.P.M.C. Schistosomiasis in the People ' s Republic of China: Prospects and Challenges for the 21st Century. *Clin. Microbiol. Rev.* **2001**, *14*, 270–295. [CrossRef]

11. Shi, F.; Zhang, Y.; Ye, P.; Lin, J.; Cai, Y.; Shen, W.; Bickle, Q.D.; Taylor, M.G. Laboratory and field evaluation of Schistosoma japonicum DNA vaccines in sheep and water buffalo in China. *Vaccine* **2001**, *20*, 462–467. [CrossRef]

12. Blas, B.L.; Rosales, M.I.; Lipayon, I.L.; Yasuraoka, K.; Matsuda, H.; Hayashi, M. The schistosomiasis problem in the Philippines: A review. *Parasitol. Int.* **2004**, *53*, 127–134. [CrossRef] [PubMed]

13. McGarvey, S.T.; Carabin, H.; Balolong, E., Jr.; Bélisle, P.; Fernandez, T.; Joseph, L.; Tallo, V.; Gonzales, R.; Tarafder, M.R.; Alday, P.; et al. Cross-Sectional associations between intensity of animal and human infection with Schistosoma japonicum in Western Samar province, Philippines. *Bull. World Health Organ.* **2006**, *84*, 446–452. [CrossRef] [PubMed]

14. Conlan, J.V.; Sripa, B.; Attwood, S.; Newton, P.N. A review of parasitic zoonoses in a changing Southeast Asia. *Vet. Parasitol.* **2011**, *182*, 22–40. [CrossRef] [PubMed]

15. Wang, T.P.; Maria, V.J.; Zhang, S.Q.; Wang, F.F.; Wu, W.D.; Zhang, G.H.; Pan, X.P.; Ju, Y.; Niels, Ø. Transmission of Schistosoma japonicum by humans and domestic animals in the Yangtze River valley, Anhui province, China. *Acta Trop.* **2005**, *96*, 198–204. [CrossRef] [PubMed]

16. Ohmae, H.; Iwanaga, Y.; Nara, T.; Matsuda, H.; Yasuraoka, K. Biological characteristics and control of intermediate snail host of Schistosoma japonicum. *Parasitol. Int.* **2003**, *52*, 409–417. [CrossRef]

17. Tanaka, H.; Tsuji, M. From discovery to eradication of schistosomiasis in Japan: 1847–1996. *Int. J. Parasitol.* **1997**, *27*, 1465–1480. [CrossRef]

18. Minggang, C.; Zheng, F. Schistosomiasis control in China. *Parasitol. Int.* **1999**, *48*, 11–19. [CrossRef]

19. Chua, C.J.; Tabios, I.K.; Tamayo, P.G.; Leonardo, L.; Fontanilla, I.K.C.; De Chavez, E.R.C.; Agatsuma, T.; Kikuchi, M.; Kato-Hayashi, N.; Chigusa, Y. Genetic Comparison of Oncomelania hupensis quadrasi Genetic Comparison of Oncomelania hupensis quad rasi (Möllendorf, 1895) (Gastropoda: Pomatiopsidae), the Intermediate Host of Schistosoma japonicum in the Philippines, Based on 16S Ribosomal RNA Sequence. *Sci. Diliman* **2017**, *292*, 32–50.

20. Leonardo, L.; Rivera, P.; Saniel, O.; Solon, J.A.; Chigusa, Y.; Villacorte, E.; Chua, J.C.; Moendeg, K.; Manalo, D.; Crisostomo, B.; et al. New endemic foci of schistosomiasis infections in the Philippines. *Acta Trop.* **2015**, *141*, 354–360. [CrossRef]

21. Pesigan, T.P.; Hairston, N.G.; Jauregui, J.J.; Garcia, E.G.; Santos, A.T.; Santos, B.C.; Besa, A.A. Studies on Schistosoma japonicum infection in the Philippines: 2. The molluscan host. *Bull. World Health Organ.* **1958**, *18*, 481. [PubMed]

22. Nihei, N.; Kanazawa, T.; Blas, B.L.; Saitoh, Y.; Itagaki, H.; Pangilinan, R.; Matsuda, H.; Yasuraoka, K. Soil factors influencing the distribution of Oncomelania quadrasi, the intermediate host of Schistosoma japonicum, on Bohol Island, Philippines. *Ann. Trop. Med. Parasitol.* **1998**, *92*, 699–710. [CrossRef] [PubMed]

23. Mcmullen, D.B. The Control Of Schistosomiasis japonica: I. Observations on the habits, ecology and Life Cycle of Oncomelania quadrasi, The Molluscan Intermediate Host Of Schistosoma japonicum in the Philippine Islands. *Am. J. Epidemiol.* **1947**, *45*, 259–273. [CrossRef]

24. Pm, B.; Ingalls, J.W., Jr. The molluscan intermediate host and schistosomiasis japonica; observations on the production and rate of emergence of cercariae of Schistosoma japonicum from the molluscan intermediate host, Oncomelania quadrasi. *Am. J. Trop. Med. Hyg.* **1948**, *28*, 567–575.

25. Belizario, V.Y.; Martinez, R.M.; de Leon, W.U.; Esparar, D.G.; Navarro, J.R.P.; Villar, L.C.; Sunico, L.S.; Velasco, L.R.; Sison, S.M. Cagayan Valley: A newly described endemic focus for schistosomiasis japonicum in the Philippines. *Philipp. J. Intern. Med.* **2005**, *43*, 117–122.

26. Leonardo, L.R.; Acosta, L.P.; Olveda, R.M.; Aligui, G.D.L. Difficulties and strategies in the control of schistosomiasis in the Philippines. *Acta Trop.* **2002**, *82*, 295–299. [CrossRef]

27. Fornillos, R.J.C.; Fontanilla, I.K.C.; Chigusa, Y.; Kikuchi, M.; Kirinoki, M.; Kato-Hayashi, N.; Kawazu, S.; Angeles, J.M.; Tabios, I.K.; Moendeg, K.; et al. Infection rate of Schistosoma japonicum in the snail Oncomelania hupensis quadrasi in endemic villages in the Philippines: Need for snail surveillance technique o Title. *Trop. Biomed.* **2019**, *36*, 402–411.

28. Driscoll, A.J.; Kyle, J.L.; Remais, J. Development of a novel PCR assay capable of detecting a single Schistosoma japonicum cercaria recovered from Oncomelania hupensis. *Parasitology* **2005**, *131*, 497–500. [CrossRef]

29. Hung, Y.W.; Remais, J. Quantitative detection of Schistosoma japonicum cercariae in water by real-Time PCR. *PLoS Negl. Trop. Dis.* **2008**, *2*, 337. [CrossRef]

30. Sato, M.O.; Rafalimanantsoa, A.; Ramarokoto, C.; Rahetilahy, A.M.; Ravoniarimbinina, P.; Kawai, S.; Minamoto, T.; Sato, M.; Kirinoki, M.; Rasolofo, V.; et al. Usefulness of environmental DNA for detecting *Schistosoma mansoni* occurrence sites in Madagascar. *Int. J. Infect. Dis.* **2018**, *76*, 130–136. [CrossRef]

31. Sengupta, M.E.; Hellström, M.; Kariuki, H.C.; Olsen, A.; Thomsen, P.F.; Mejer, H.; Willerslev, E.; Mwanje, M.T.; Madsen, H.; Kristensen, T.K.; et al. Environmental DNA for improved detection and environmental surveillance of schistosomiasis. *Proc. Natl. Acad. Sci. USA* **2019**, *116*, 8931–8940. [CrossRef] [PubMed]

32. Google Earth. 2019. Available online: https://earth.google.com/web/@18.27337328,122.10486135,232.93976988a,88309.32932935d,35y,0h,0t,0r/data=ChMaEQoJL20vMDZqX2hzGAIgASgCNoTitle (accessed on 31 July 2019).

33. Wilfinger, W.W.; Mackey, K.; Chomczynski, P. Effect of pH and ionic strength on the spectrophotometric assessment of nucleic acid purity. *Biotechniques* **1997**, *22*, 474–481. [CrossRef] [PubMed]

34. Desjardins, P.; Conklin, D. NanoDrop microvolume quantitation of nucleic acids. *JoVE* **2010**, *45*, 2565. [CrossRef] [PubMed]

35. Schneider, J.; Valentini, A.; Dejean, T.; Montarsi, F.; Taberlet, P.; Glaizot, O.; Fumagalli, L. Detection of invasive mosquito vectors using environmental DNA (eDNA) from water samples. *PLoS ONE* **2016**, *11*, e0162493. [CrossRef] [PubMed]

36. Buxton, A.S.; Groombridge, J.J.; Griffiths, R.A. Seasonal variation in environmental DNA detection in sediment and water samples. *PLoS ONE* **2018**, *13*, e0191737. [CrossRef] [PubMed]

37. Ellison, S.L.R.; English, C.A.; Burns, M.J.; Keer, J.T. Routes to improving the reliability of low level DNA analysis using real-time PCR. *BMC Biotechnol.* **2006**, *6*, 33. [CrossRef]

38. Bohmann, K.; Evans, A.; Gilbert, M.T.P.; Carvalho, G.R.; Creer, S.; Knapp, M.; Douglas, W.Y.; De Bruyn, M. Environmental DNA for wildlife biology and biodiversity monitoring. *Trends Ecol. Evol.* **2014**, *29*, 358–367. [CrossRef]

39. Herder, J.; Valentini, A.; Bellemain, E.; Dejean, T.; Van Delft, J.; Thomsen, P.F.; Taberlet, P. *Environmental DNA—A Review of the Possible Applications for the Detection of (Invasive) Species*; Technical Report: Report number: 2013-104; RAVON: Amsterdam, The Netherlands, 2014.

40. Rees, H.C.; Maddison, B.C.; Middleditch, D.J.; Patmore, J.R.M.; Gough, K.C. The detection of aquatic animal species using environmental DNA—a review of eDNA as a survey tool in ecology. *J. Appl. Ecol.* **2014**, *51*, 1450–1459. [CrossRef]

41. Thomsen, P.F.; Willerslev, E. Environmental DNA—An emerging tool in conservation for monitoring past and present biodiversity. *Biol. Conserv.* **2015**, *183*, 4–18. [CrossRef]

42. Goldberg, C.S.; Turner, C.R.; Deiner, K.; Klymus, K.E.; Thomsen, P.F.; Murphy, M.A.; Spear, S.F.; McKee, A.; Oyler-McCance, S.J.; Cornman, R.S.; et al. Critical considerations for the application of environmental DNA methods to detect aquatic species. *Methods Ecol. Evol.* **2016**, *7*, 1299–1307. [CrossRef]

43. Thomsen, P.F.; Kielgast, J.; Iversen, L.L.; Møller, P.R.; Rasmussen, M.; Willerslev, E. Detection of a diverse marine fish fauna using environmental DNA from seawater samples. *PLoS ONE* **2012**, *7*, e41732. [CrossRef] [PubMed]

44. Eichmiller, J.J.; Best, S.E.; Sorensen, P.W. Effects of temperature and trophic state on degradation of environmental DNA in lake water. *Environ. Sci. Technol.* **2016**, *50*, 1859–1867. [CrossRef] [PubMed]

45. Johnson, P.T.J.; Chase, J.M.; Dosch, K.L.; Hartson, R.B.; Gross, J.A.; Larson, D.J.; Sutherland, D.R.; Carpenter, S.R. Aquatic eutrophication promotes pathogenic infection in amphibians. *Proc. Natl. Acad. Sci. USA* **2007**, *104*, 15781–15786. [CrossRef] [PubMed]

46. Odlare, M.; Arthurson, V.; Pell, M.; Svensson, K.; Nehrenheim, E.; Abubaker, J. Land application of organic waste—Effects on the soil ecosystem. *Appl. Energy* **2011**, *88*, 2210–2218. [CrossRef]

47. Khan, S.A.; Mulvaney, R.L.; Ellsworth, T.R. The potassium paradox: Implications for soil fertility, crop production and human health. *Renew. Agric. Food Syst.* **2014**, *29*, 3–27. [CrossRef]

48. Shuhaimi-Othman, M.; Nur-Amalina, R.; Nadzifah, Y. Toxicity of metals to a freshwater snail, Melanoides tuberculata. *Sci. World J.* **2012**, *2012*, 125785. [CrossRef]

49. Gawad, S.S.A. Acute toxicity of some heavy metals to the fresh water snail, Theodoxus niloticus (Reeve, 1856). *Egypt. J. Aquat. Res.* **2018**, *44*, 83–87. [CrossRef]

50. Moreno, L.I.; McCord, B.R. Understanding metal inhibition: The effect of copper (Cu^{2+}) on DNA containing samples. *Forensic Chem.* **2017**, *4*, 89–95. [CrossRef]

51. Cervantes-Cervantes, M.P.; Calderón-Salinas, J.V.; Albores, A.; Muñoz-Sánchez, J.L. Copper increases the damage to DNA and proteins caused by reactive oxygen species. *Biol. Trace Elem. Res.* **2005**, *103*, 229–248. [CrossRef]

52. Jove, J.A. Use of molluscicides in the control of bilharziasis in Venezuela: Equipment and methods of application. *Bull. World Health Organ.* **1956**, *14*, 617.

53. Chandiwana, S.K.; Ndamba, J.; Makura, O.; Taylor, P. Field evaluation of controlled release copper glass as a molluscicide in snail control. *Trans. R. Soc. Trop. Med. Hyg.* **1987**, *81*, 952–955. [CrossRef]

54. Fahmy, S.R.; Abdel-Ghaffar, F.; Bakry, F.A.; Sayed, D.A. Ecotoxicological effect of sublethal exposure to zinc oxide nanoparticles on freshwater snail Biomphalaria alexandrina. *Arch. Environ. Contam. Toxicol.* **2014**, *67*, 192–202. [CrossRef] [PubMed]

55. Ge, J.Y.; Yang, X.H.; Yang, Y.; Feng, H.H.; Hu, X.Y.; Wang, X.Y. The Facile Preparation and the Performance of Killing Oncomelania Snails of Zinc Oxide Nanoparticles. *Adv. Mater. Res.* **2014**, *1078*, 40–44. [CrossRef]

56. Bolyard, S. Fate of Zinc Oxide in Landfill Leachate. Master's Thesis, University of Central Florida, Orlando, FL, USA, 2012.

57. Zhou, X.N.; Wang, L.Y.; Chen, M.G.; Wu, X.H.; Jiang, Q.W.; Chen, X.Y.; Zheng, J.; Jürg, U. The public health significance and control of schistosomiasis in China—Then and now. *Acta Trop.* **2005**, *96*, 97–105. [CrossRef] [PubMed]
58. Mänd, R.; Tilgar, V.; Leivits, A. Calcium, snails, and birds: A case study. *Web Ecol.* **2000**, *1*, 63–69. [CrossRef]

MDPI

St. Alban-Anlage 66

4052 Basel

Switzerland

Tel. +41 61 683 77 34

Fax +41 61 302 89 18

www.mdpi.com

Pathogens Editorial Office

E-mail: pathogens@mdpi.com

www.mdpi.com/journal/pathogens